# Measuring Workplace Performance

## Second Edition

# Measuring Workplace Performance

## Second Edition

Michael J. O'Neill

**CRC Press**
Taylor & Francis Group
Boca Raton  London  New York

CRC Press is an imprint of the
Taylor & Francis Group, an **informa** business
A TAYLOR & FRANCIS BOOK

CRC Press
Taylor & Francis Group
6000 Broken Sound Parkway NW, Suite 300
Boca Raton, FL 33487-2742

First issued in paperback 2019

ISBN-13: 978-0-8493-5801-2 (hbk)
ISBN-13: 978-0-367-39044-0 (pbk)

---

**Library of Congress Cataloging-in-Publication Data**

---

O'Neill, Michael J., 1959-
    Measuring workplace performance / Michael J. O'Neill. -- 2nd ed.
      p. cm.
    Rev. ed. of: Ergonomic design for organizational effectiveness. 1998.
    Includes bibliographical references and index.
    ISBN-13: 978-0-8493-5801-2 (alk. paper)
    1. Human engineering. 2. Labor productivity. 3. Human beings--Effect of environment on. 4. Work environment. 5. System design. I. O'Neill, Michael J., 1959- Ergonomic design for organizational effectiveness. II. Title.

TA166.O54 2006
620.8'2--dc22                                    2006013568

---

**Visit the Taylor & Francis Web site at**
**http://www.taylorandfrancis.com**

**and the CRC Press Web site at**
**http://www.crcpress.com**

# Acknowledgments

For my wife Danelle O'Neill and son William O'Neill.
Thank you for your love and support.

As is true with any body of work conducted over a period of years, many people have contributed in different ways to my thinking, research and consulting. Patricia Bergquist made a primary contribution by being an important part of much of the original research, and many of the Case Studies presented in this book. She along with Yvonne Boucher and Julie Sless extended my thinking on the use of business metrics in workplace research.

Others have supported this program in various ways, including Michael Volkema, Brian Walker, Kris Manos, Mark Kinsler, Sheryl Smith, Lois Maassen, Joanie Reid, Judy Leese and the OERC, and many other colleagues within Herman Miller, Inc. I am deeply indebted to the many people who took the time to read early drafts of this manuscript and provide invaluable feedback and support in other ways, especially Clark Malcolm, Brian Green, Rick Marken, Jim Long, William O'Neill, and Larry Scheerer. Special thanks to Stuart Hamilton for designing the cover of this book. My editor at Taylor & Francis, Cindy Carelli, made this book possible with her help on several key issues.

Finally, there would be little to write about if it were not for the companies that I have had the privilege to consult with over the years, and the individuals within those organizations who have valued this work, and sustained this program.

To all these people, my sincere thanks and appreciation.

# The Author

**– Dr. Michael J. O'Neill**

Dr. O'Neill leads the Workplace Performance Metrics practice area within the Herman Miller, Inc. Services group. He has 18 years' experience in conducting research projects for Fortune 1000 companies that assess the impact of work environment design on behavioral and business outcomes. A Certified Six Sigma Master Black Belt, he also specializes in implementing quality measures programs related to workplace design and space management. Michael is a Board Certified Professional Ergonomist with a Ph.D. in Architecture, Master of Architecture, and a B.A. in Psychology.

He speaks internationally and has written a book, *Ergonomic Design for Organizational Effectiveness* (1998), is a co-author of the *BSR/HFES100 HFES Computer Standards* (2005), numerous book chapters, and over 30 articles on workplace design and human performance.

Michael has conducted workplace research and consulting projects for companies in the Agriculture, Automotive, Consulting, Consumer Products, Energy, Financial Services, and Telecom industries in the US, England, Canada, Europe, and Asia.

# INTRODUCTION

In this Second Edition of *Measuring Workplace Performance*, we not only provide the reader with state of the art theory, research and methods, but have made every effort to stress clarity of writing and in expression of ideas. In terms of new content, we have added 10 new Case Studies (for a total of 17 in the book), with 60 new Tables and 60 new Figures. The presentation of concepts and information within the book has been re-ordered and streamlined, to enhance understanding. Every existing Chapter has been extensively rewritten with an emphasis on eliminating technical "jargon," so material is accessible to a wide range of readers. We hope that you find this updated edition to be useful and thought provoking.

An organization's workplace design strategy has far-reaching effects (good and bad) on internal culture, retention, attraction, and the health and performance of employees. Some organizations follow a workplace strategy that emphasizes cost reduction, or ease of facility management. These organizations have a point of view that the physical workplace does not influence performance or business effectiveness. Unfortunately, these companies miss the opportunity to use workplace design to address business objectives related to creating effective workplaces, such as: using the workplace to enhance sense of community in employees, to reflect corporate "brand," to increase collaboration, communication, innovation, or to increase the speed and efficiency of business processes. Some might argue that we can't "prove" that the physical workplace affects performance, so why invest? This book illustrates that we can measure and show credible links between workplace design features, and human performance and business outcomes.

Companies regularly invest in technology and employee development programs in the implicit belief that some of this investment will translate into competitive advantage. Similarly, the facility and workplace is an additional "lever" that management can pull to enhance performance. The challenge to organizations is to design and manage facilities against the dynamic, moving target of business strategy and tactical requirements. A further challenge is to somehow measure the performance of facilities in terms of their impact on work performance of employees.

To address this challenge, we offer a dynamic framework for understanding organizations and their physical workplaces, and an ongoing measurement methodology to analyze workplace performance. Thus, the focus of this book is on measuring the alignment between the physical office work environment, and human performance and business objectives.

As part of the dynamic framework, we employ a "biological metaphor" to understand the function of work organizations and in particular, physical workplaces. The idea of a biological model for understanding phenomena has been applied over the years to areas such as human cognition (Kaplan and Kaplan, 1989; O'Neill, 1991) economics (see Rothschild's 1990 "Bionomics" book), technology and business (Frenay, 2006). The biological model has recently been applied to understanding the transactions between the organization and its physical workspace (O'Neill, 1998). We extend this framework in the current volume.

A key premise of our biological metaphor is that *environmental control* is the dynamic mechanism by which the physical workspace can be adapted and aligned to meet the purpose of the organization. Further, environmental control can be implemented at different levels of the physical and social organization -- at the level of the organization/business unit, the group, and the individual.

Throughout the book we show that measuring the impact of workspace design on specific business outcomes (both human performance and financial) is critical to the successful implementation, ongoing management and improvement of office work environments. To this end, we present a measurement model and methods based on six-sigma approaches and tools.

## I. POINT OF VIEW

This volume presents a conceptual model for thinking about the physical, technical and social components of organizations, and the internal processes and external forces that drive change in them. A central theme: workplace design that enhances *control over the physical environment* is a critical mechanism for supporting ever-changing shifts in organizational goals and structure. Environmental control is the means by which the system optimizes the form of the environment in support of the behaviors needed to meet business goals. Increased control over the work process and work environment has consistently been shown to enhance the health and effectiveness of workers and the organization (Karasek and Theorell, 1990).

In this book, we examine organizations and work spaces using the metaphor of a biological system. The system consists of social (people, organizational structure), technical (machines, information technology, rules of business), and environmental (physical work place) processes or components. These components interact (more or less effectively) in the pursuit of attaining business goals. Control over the physical environment is a key mechanism that can be "designed in" to optimize the form of the work environment and ultimately support organizational effectiveness. Workplace design should explicitly support the *purpose* of the organization (as opposed to design for design's sake).

In this book, we focus on "white collar" or "professional" work that takes place in office settings. We discuss the tools and methods that we have applied to understand and predict ever-changing workplace design requirements for organizations. Central to this book is the application of effective measurement methods that can link human performance and business outcomes with physical workplace design features.

## II. ORGANIZATION OF THE BOOK

*Measuring Workplace Performance* is divided into three parts. Part I, "The Organization and Workplace as a Biological System" describes key components of the biological system as a metaphor for understanding the function of organizations and the physical workplace. Chapter 1 discusses competing office workplace metaphors. Chapter 2 describes the Biological Systems model. Chapter 3 discusses how Environmental Control, which is a key mechanism for dynamics and change within the Biological Model, can be applied through workplace design to improve health, performance, and effective work.

Part II, "The Workplace Performance Measurement Model," focuses on methods to create, manage and measure the performance of work environments. In Chapter 4, we discuss the process of workplace measurement within a Quality framework.

Part III, "Case Studies: Facility/Building, Group, and Individual Spaces" contains the Case Studies in which workplace performance is measured at the organization, group/team, and individual levels. Chapter 5 discusses five Case Studies at the facility/organization level. Chapter 6 describes five Case Studies at the group/departmental level of analysis. Chapter 7 discusses five Case Studies at the individual workspace level. These Case Studies explore the relationship between work environment design (including environmental control) and various behavioral and financial outcome measures, along with observations about the results.

## III. CONCLUSION

My goal is not to advocate for a particular office workplace design solution, but to illustrate the application of the biological model for organizations and workspaces, and the use of our workplace measurement model. We use the Case Studies to show how environmental control has been employed at different scales of the organization and workspace to enhance performance.

Individuals at all levels of the organization, from finance and human resources to real estate and facilities management, have a say in shaping our work environments. My hope is that by providing an inclusive framework to define and measure the impact of workplace design, it will help to forge a new mind-set about the role the workplace can play in improving organizational performance.

# Contents

# PART I: THE ORGANIZATION AND WORKPLACE AS A BIOLOGICAL SYSTEM

CHAPTER 1

# Workplace: Machine or Living Entity?

Most people, including business leaders and professionals engaged in the design or management of office work environments, assume certain "givens" about the way the world operates and then act in accordance with that belief system. This belief system can significantly affect the way office work environments are designed and implemented. The work environment, in turn, affects the behavior and performance of employees who use those spaces and, to some degree, the success of organizations. In this chapter we contrast "machine" and "biological" metaphors for the way people understand the world. We then explore how these metaphors have been (and could be) applied to the design of organizations and office workspaces.

## I. MACHINE VERSUS BIOLOGICAL METAPHORS

While a number of belief systems filter the way we interpret or predict events in the world, the machine metaphor has been responsible for driving enormous change in technology, culture, and human relations in the past century. The biological metaphor is currently emerging as a much better way of understanding phenomena in various fields, like cognitive science and organizational behavior (Kaplan and Kaplan, 1989; Land and Jarman, 1993; O'Neill, 1991b, 2005).

These metaphors are most visible in the physical form taken by buildings and office workplaces -- which are reflections of the metaphors (whether recognized or not) that influenced the business organizations that built them. In this chapter, we describe and contrast "machine" and the "biological" metaphors for understanding the world, and their impact on organizations and the design of office workspaces.

Table 1.1 provides descriptors to illustrate the contrast between the belief systems, and to provide a basis later in this discussion for thinking about how each of them might influence the design and development of office workplaces.

### A. The Machine Metaphor

The metaphor of the machine -- gears, levers, springs, circuits, control mechanisms, and related assumptions about the way a machine functions has had a powerful influence on how people interpret events that occur in the world around them.

## 1. Reductionism

A central characteristic of the machine metaphor is that components of any problem, event or phenomenon, like those of a machine, can be broken into discrete parts, analyzed, designed, and its activities examined.

| Table 1.1  Contrasts Between Belief Systems About the World ||
|---|---|
| **Machine Metaphor** | **Biological Metaphor** |
| Reductionism: Phenomenon can be broken into discrete separate entities, events, and examined | System-level analysis -- system cannot be reduced to individual components and studied |
| Individual as unit of analysis | Group, units of organizations, systems, as units of analysis |
| Cause and effect relationships between events | No direct "cause and effect" -- Transactions between subsystems can change form and behaviors of organization and workplace |
| Control mechanism required to manage operation of system components | Self-managing behavior |
| Independent observer | Observer is part of phenomena |

## 2. Individual as Unit of Analysis

When studying human or organizational behavior (or designing organizations and workplaces), the unit of analysis is the individual (or a discrete piece of an event).

## 3. Cause and Effect

This analogy uses a "cause and effect" model of relationships that is often applied to predicting, understanding, or rationalizing events in the world. Sequences of events are often seen as being orderly and moving in a specific direction or flow, without considering other factors that may be influencing outcomes.

## 4. Control Mechanism

Within this metaphor, a control function of some sort is required to organize, coordinate, and manage the activities of the components of the system.

The machine metaphor requires a "homunculus" of some sort to guide the operation of its components (Kaplan and Kaplan, 1989).

## 5. Independent Observer

The notion of the independent observer states that the observer is separate from, independent of, and does not influence the phenomena being examined. In the machine metaphor, the observer can stand apart from the phenomena being studied and objectively observe and measure events without affecting the results.

## B. The Biological Metaphor

In the biological (sometimes also referred to as a "natural system") metaphor, the organization is a dynamic *system* within which people, technology, process, and the environment form *subsystems*, each actively influencing the other (Altman and Rogoff, 1987).

## 1. System Level Analysis

Unlike the reductionism of the machine metaphor, entire (sub) systems are the unit of understanding - and of design. Some examples of subsystems within organizations include technical subsystems (tools and processes), social subsystems (social networks), and workspace subsystems (offices, meeting spaces, buildings). In this view it is meaningless to analyze specific "pieces" of a phenomenon taken out of the larger context of the system in which it exists. Typically group work, or processes that cut across departments or business units, is the unit of analysis, and it is not possible, nor desirable, to analyze individual work activities or outcomes piecemeal. The individual parts of an organization cannot be studied, or designed, in isolation from each other.

In this metaphor, the subsystems (technology, process, environment) making up the larger whole are *subordinate to the larger purpose of the system* (Kitchener, 1982). This also suggests the potential for purposeful design of an organization (and the office environment) with the goals of the business in mind. In this metaphor, the workplace is designed to meet a specific purpose or objective (see Table 1.1).

## 2. Transactions between Subsystems

In the natural system, one subsystem may affect another subsystem, a principle known as "efficient causation" (Altman and Rogoff, 1987). This is not the same as the deterministic "cause and effect" characteristic of the machine metaphor, in which one event triggers the next - like billiard balls on a pool table (see Table 1.1). Rather, "efficient causation" is related to learning and adaptive behavior. Subsystems (such as the office workspace, or technology) can be designed to take in information from other subsystems, easily adapting new forms, or behaviors, or capabilities in reaction to learning or feedback from the other parts, and thus enhance the flexibility of the overall organization.

### 3. Self-Managing Behavior

The form or configuration of these subsystems is self-managed by the subsystems themselves; they can easily change over time in response to internal and external forces. The ability to learn, change, and grow is built into the structure of the sub-systems themselves. A separate controlling function (homunculus) as found in the machine analogy is not required for the system to work.

### 4. Observer is Part of the System

In the biological framework, the observer (for instance, the designer of the system or researcher making observations) is by definition a participant in the system itself. The observer cannot stand separate from the system; the act of observation itself influences the behavior of the system.

## II. APPLICATION OF THE MACHINE AND BIOLOGICAL METAPHORS TO THE WORKPLACE

The machine metaphor has been widely explored in architecture, most notably through the work of architects such as Le Corbusier and Gropius. An examination of many existing office spaces suggests that the machine analogy continues to dominate the world of interior office design.

Aspects of the biological metaphor have been embraced in several areas of science and business, most notably in psychology, organization development, and knowledge management. In the area of office workplace design and management, the idea of applying biological metaphor concepts is being explored by leading organizations.

Table 1.2 provides a summary of descriptors that contrast characteristics of workplaces using the machine and biological metaphors.

| Table 1.2 Comparison of Machine and Biological Metaphors for Workspaces ||
|---|---|
| **Machine Metaphor** | **Biological Metaphor** |
| Workplace is an unavoidable overhead cost | Workplace as asset: A tool for effective work |
| Environment not linked to business strategy, may reflect hierarchy or other issues | Workplace designed to support business objectives, mission |
| Individuals have limited control over workspace | Individuals, groups and departments have control over workspaces |
| Control: Static, not flexible. Design reinforces order, reacts to current problems | Control: Accommodates change. Dynamic, flexible. Design anticipates future needs |
| Viewpoint: Individual. Emphasize individual workspace, individual activities | Viewpoint: Enterprise. Space supports collaboration between people and groups, flow of business processes |

## A. The Machine Metaphor and the Workplace

This section discusses general characteristics of organizations and work environments designed from the perspectives of the machine metaphor.

### 1. Unavoidable Overhead Cost

The environment as machine is viewed purely as an overhead cost to the organization, not as a potential tool for strategic advantage. Under this model, Real Estate and Facility Managers are under constant pressure to reduce these overhead costs through space efficiency and space reduction programs.

### 2. Workplace not Linked to Business Strategy

Applied to the development of organizational design and workplace strategy, the assumption is that work processes are entirely predictable and are biased towards individual work. Thus workspaces are designed to support individual activities. Support for group work, business processes or organizational objectives are not addressed by workplace design.

Workspaces are often designed to indicate the individual's status level within the organization. Reflection of hierarchy is important to the machine metaphor because attached to status are specific, static, roles and norms. The predictability of roles, norms, and responsibilities is important to the smooth functioning of the parts within the machine. Of course this approach to design has little relationship to supporting specific goals of the mission of the organization. Rather, the machine approach is directed internally, to the smooth function of the machine. When an organization is designed (intentionally or otherwise) according to the machine metaphor, the approach works well as long as the overall organization remains aligned properly with its external environment. When the external (business) environment changes, organizations using this model grasp ever more rigidly to the rules and roles of its internal functions, including the design of the organization and the physical environment. Thus the parts of the organization, including the office workplace, get out of alignment with the business mission when change occurs, because ability to accommodate change and align with business objectives is not inherently built into the system.

## 3. Limited Control over Workspace

The machine analogy, as applied to the work environment, results in an emphasis on individual workspaces with limited adjustability. It is not required to give individuals control over their workspace (through adjustability of components) because the design has already been closely optimized to support a specific set of highly defined work processes. The design strategy is not intended to support unanticipated changes in work activities, workflow, or changes in business requirements. Only the designers of the "machine" have enough knowledge to make a change to its design.

## 4. Control: Static, not Flexible, Design Reinforces Order

The workspace is not designed with the intention of supporting change. When organizational change does occur, however, it is very disruptive because the physical workspace lags behind in terms of its ability to support new ways of working. The workplace in this metaphor is a static mechanism, supporting order within the system, maintaining the system as it was originally designed. It is difficult to modify or change the design of the mechanism (the office environment and supporting technology) to address new needs. The design of the mechanism reacts at best to current needs, most often to past ones, and never accounts for the future.

In this machine analogy, time, context (location or space), and change are not directly considered. Time and location are not considered as part of the functioning of the machine.

Within this analogy *behavior is under the control of the environment.* The implication of this view is that it is possible to design the environment to cause people to behave in specific, predictable ways -- and thus support specific work processes. This viewpoint is useful for designing and managing organized manual work, such as work on assembly lines, in which the work process is often linear, the emphasis is on the individual, and productivity is assessed in terms of quantifiable output, such as number of objects created per unit of time, or number of operations performed on objects in a production setting.

5. Viewpoint: Individual

The machine analogy focuses on individual work activities, processes and places for work to occur. Work processes are highly defined, isolated, and proscribed. Work processes are replicable so that any worker with reasonable training can perform them. Within an organization, all business processes and functions are also highly defined so that overall, work activities are predictable. Thus in such a design it is possible to have all individual work activities interact in a predictable manner, like gears meshing within a complex machine (see Figure 1.1). An overall control mechanism (the Management function) monitors the individual activities and keeps them in sync. The design of the system is fixed, and there is limited capability to adapt to new situations and change from external forces (Pepper, 1942).

When applied to office design, a mechanistic orientation toward the work environment suggests that the individual parts of the workstation, such as the computer, desk, and seating, can be independently considered, and that there will be interplay between these individual elements.

**Figure 1.1** Illustration of the machine metaphor (Author)

The machine analogy has also been applied to the traditional way in which Call Center work and other "back room" business operations, including the workplaces to support them, have been designed (see Figure 1.2). Figure 1.2 shows a typical Call Center workstation designed using the Machine Metaphor. In this analogy, individual workspaces might be centralized within a single, large contiguous building space. The layout of workstations could be designed using a large grid of low height cubicle walls for ease of visual monitoring and tracking of location of employees.

The technical system might be designed to include electronic performance monitoring programs in which customer phone calls are randomly monitored and employee performance evaluated. Processes for interacting with customers could be highly specified, including scripts that operators read for different situations. The role of management in such a system is to ensure the processes are followed, thus the analogy of gears within a machine being kept "in sync" by management (Figure 1.1).

6. Observations about the Machine Metaphor

In terms of supporting business needs, the companies employing this metaphor view the workplace as an unavoidable cost of doing business, rather than a strategic investment that can create competitive advantage. The shortcomings of this approach abound, including its static design, focus on the individual as the unit of analysis, failure to consider group or team work, lack of control over the workspace, and the failure to accommodate organizational change.

**Figure 1.2** Call Center agent's workstation -- Machine Metaphor (Author)

A common problem in designing with the machine metaphor is that individual aspects of the system (such as workstation standards) are addressed independently of each other. The designers fail to consider the larger system, such as relationships between business units, informal social networks, and other aspects that should also be incorporated into a successful design solution.

The machine metaphor is most appropriate when applied to situations in which work processes are well defined and repeatable (such as certain types of data processing or assembly work), and outcomes are clearly quantifiable (such as piece count per unit of time). When this worldview is allowed to influence the design of environments for most other types of workers, particularly knowledge workers, the results are predictable. Frequent complaints about and frustrations with this viewpoint include: lack of support for group or team activities; poor response to technology drivers; design lacks a true business context; the design has no apparent link to broader issues of organizational effectiveness; and general lack of flexibility in the work environment to accommodate change due to internal restructuring or new business opportunities.

## B. The Biological Metaphor and the Workspace

The biological metaphor is one of living things, natural systems, and the living organism (see Figure 1.3).

**Figure 1.3** Nature -- the Biological Metaphor (Author)

1. Workplace as an Asset

The previous discussion about the machine metaphor illustrates an important contrast between that and the biological metaphor - facility as liability versus asset (Vischer, 1996). Cost considerations exert considerable pressure on planning for the accommodation of workers. As companies reorganize, whether enlarging or reducing their work forces, the costs associated with housing employees and providing their work tools continues to increase. The metaphor chosen for the workspace influences the economic perspective that an organization has on the role of the workspace in business (see Table 1.2). Managers using the machine metaphor typically view the facility as purely a cost center. The facility strategy with this metaphor will be marked by reductions in owned and leased space, reduced service amenities, deferred building maintenance programs, and other results associated with reduced budgets.

This cost perspective can ultimately lead to a reduction in the quality of the work environment and, we believe, in the potential contribution that the environment can make to organizational effectiveness. Because real estate and facilities costs are such an obvious target, programs around reducing these obvious costs are often implemented first, before more difficult business decisions have to be made.

Alternatively, the biological metaphor suggests that the workplace is an investment made by the organization to enhance performance and to fully integrate the facility with the business mission. Thus the workplace may not be designed primarily to reduce space or cost of space, but to support the work style, business objectives, and to convey the culture and values of the organization. In this perspective, the workspace is designed as part of a strategy that carries the expectation that the work environment will support the work process and, in turn, the creation of value to the organization.

## 2. Accommodates Change

An important aspect of the biological metaphor is that time and change are "built in" to the system. Thus, the system is inherently capable of changing over time to adapt to changing environmental conditions or demands (see Table 1.2).

## 3. Workplace is Designed to Support Business Objectives

In the biological metaphor, the whole of the system (organization, technology, physical workplace) is given meaning *by its defined purpose* (Reese and Overton, 1970). In such a perspective, the emphasis shifts from attempting to describe how to do things (work activities and processes) to a focus on the product that results from these actions or other higher-level objectives. This "purpose" focus is common among startup companies and other smaller entrepreneurial efforts, in which roles and specific activities are de-emphasized in favor of reaching business goals. In this metaphor, the workspace is designed for a specific purpose - to attain specific organizational goals, such as behavioral change (enhanced collaboration, etc.), business process improvements, or other defined goals. Figure 1.4 shows a concept for highly adjustable workspace that can be reconfigured to support changes in work behaviors required by business objectives.

## 4. Control. Individuals and Groups Have Control over Workspaces

Within the biological metaphor, the emphasis shifts from a strategy of control over people to a strategy of providing employees with optimal control over their jobs, and the work environment. Recent research shows a growing link between enhanced control over the workspace and increased job control (see later Case Studies within this volume). A large and established body of research shows a link between increased job control and reduced risk of stress and coronary heart disease (CHD) (Karasek and Theorell, 1990).

Thus the physical work environment might be designed to support high levels of individual adjustability and work team support at the expense of visual monitoring by supervisors. A call center could become a learning environment in which teams support each other, and manage their workflow, and in which individuals learn from each other. Instead of using only individual workstations, such an environment could also include varying types of meeting spaces for different size groups in order to facilitate communication and learning. Other forms of performance measurement, such as team or business unit goals, might be implemented in place of electronic performance monitoring. Under such a model, the role of management becomes one of selecting the right skill sets, leadership development, and coaching of employees.

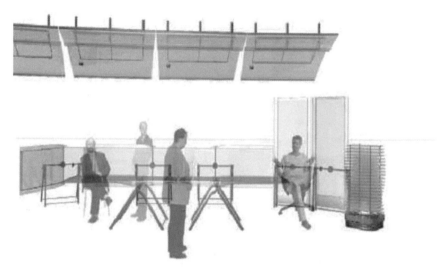

**Figure 1.4** Workspace design employing Biological Metaphor Concepts (With permission of Herman Miller, Inc., copyright 2005, All Rights Reserved.)

5. Workplace Accommodates Change

Since the nature of the output required of the business organization (products and services) can change in a relatively short period of time due to the nature of market conditions and customer demands, the overall work environment must be flexible enough to respond to those shifts. Thus, the focus of the office design process within the biological metaphor is to provide workplaces that possess the ability to change rapidly in unpredictable contexts. In the machine metaphor, work and the workplace are designed in stasis - for one point in time in the history of the organization. Neither the "pieces of the machine" nor management processes are designed for change over time or to accommodate new demands in the business environment. The biological metaphor explicitly considers time and changes in the subsystems over time in reaction to the external environment.

In a biological metaphor, the workspace is conceived as a self-regulating, flexible mix of features and capabilities that support a variety of work styles and processes, and gracefully accommodate change. This system provides workers control over their environment to dynamically react to changing needs and work processes rising from the purpose of the organization as it reacts to changing business and market conditions. Design using the biological metaphor anticipates change by incorporating flexible design concepts. It supports no single "order" of things, but can support the types of organized chaos that groups and individuals oriented toward a common purpose will create en route to that goal. This metaphor also reflects the inherently non-mechanistic nature of human beings.

Thus, the biological metaphor supports a different perspective on the provisioning of office work environments, especially in support of knowledge work, in which the work process is inherently unpredictable. In this approach, the environment must support the lack of predictability in work process, due to the shift in focus from job design and monitoring of tasks, to working towards organizational goals.

## 6. Viewpoint: Enterprise

The viewpoint of the biological metaphor is at the enterprise/organization level, which includes explicit consideration of facility design and layout issues related to business units, departments, and group spaces. The design of spaces reflects the organizational purpose and mission, as opposed to, for instance, reflecting individual status or position within an organizational hierarchy.

## 7. Observations about the Biological Metaphor

Companies employing this metaphor view the workplace as a strategic asset, an investment that can be leveraged to gain competitive advantage in the marketplace. Thus the design of office space is oriented toward achieving (and success measured on) business objectives, as opposed to compartmentalized design requirements.

The objective of this approach is to support group and team knowledge work, and business processes that flow across groups and departments. A key aspect of this metaphor is the concept of designing environmental control (through adjustability and flexibility of space) into the system at all levels, including individual workspace, group spaces, and facility design features. Control is seen as the mechanism to permit the workplace to "flex" and change as required by changes in the organization. We explore the concept of environmental control in the next chapter and throughout this volume.

# CHAPTER 2

## The Biological Systems Model

"Influencing behaviour is almost all of what management is about, and buildings influence behaviour." J. Seiler, 1984

In Chapter 1 we discussed the characteristics of the Biological Metaphor. In this metaphor the design of office work environments is aligned with the purpose, or business mission of the organization, rather than by other issues that cannot be shown to directly support business objectives. In this way, the design and function of the system are "pulled" or aligned to organizational mission and purpose. Effective work is thus a natural outcome of an organization and workplace designed to support the biological metaphor. When workplaces are not designed with this larger viewpoint in mind, the design process runs the risk of being sidetracked by issues such as: using workplace to reflect hierarchy, the inertia of existing workplace standards, or short-term cost considerations.

In order to create a viable office work environment strategy it is necessary to understand the entire organization as a system, to determine how the workplace can be designed to effectively support the organization's purpose. Thus, the creation of a workplace design does not begin with designing the features of the work space -- rather it begins with an understanding of the objectives of the group or business unit using that space. These business objectives in turn should drive the design requirements for the social, technical, and workplace subsystems that support those objectives.

In this chapter, we begin with a detailed discussion of the functions and processes of the biological metaphor. Thus the metaphor is translated into a working model that reveals the dynamic nature of the biological system and how it reflects the behavior of organizations, and workplaces, over time. The chapter concludes with a summary of characteristics of the model and how it can be applied to the workplace design process.

## I. DESCRIPTION OF THE MODEL

The components and processes of the Biological Systems model are graphically depicted in Figure 2.1. This model emphasizes the flow of input (in knowledge work, this is raw data or information), and the transformation of this information into a knowledge product that has value to a customer, and that creates value to the organization. Ultimately, this product (or output), of the system furthers the business mission of the organization. This model can be applied at the level of an entire organization, or at smaller units, such as the department of workgroup. The model is probably of most practical value when applied to the department or workgroup levels.

The model itself includes three major subsystems: the social subsystem, the technical subsystem, and the environmental subsystem (see Figure 2.1). Note that the workplace (environment) subsystem contains within it, the social and technical subsystems. This is because the social and technical aspects of work, as well as work processes, occur within the context of the physical work environment (see Figure 2.1).

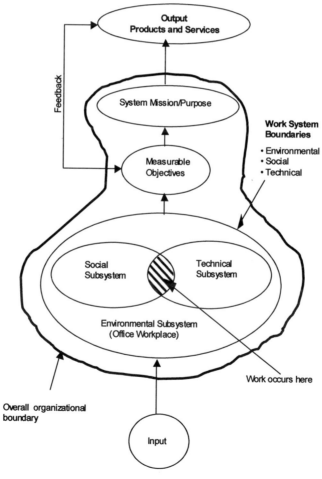

Figure 2.1 The Biological Systems Model (Author)

## A. Elements of the Biological Systems Model

### 1. System Mission/Purpose

A critical element of the model "purpose" of the organization (see Figure 2.1). In business organizations, the "purpose" is the business mission. Because the biological model is a *purposeful* system, its main activity is to transform "inputs" to the system, into "output," the output being products or services that generate economic value to the organization, its shareholders and customers. This output is aligned with the purpose. The purpose of the system will also influence organizational culture, values, and other characteristics.

The work environment can be designed as a means of achieving the purpose of the organization. In our discussion of the work environment throughout this volume, we frequently express the notion that the work environment can be used as a tool to support the work effort toward a specific purpose. Thus the design criteria or guidelines for the work environment must align with the organizational purpose, and must be viewed as being open to continuous change.

Change will surely appear from one of a number of business change drivers (discussed in a following section of this chapter). In general, change to organizations can be driven by political, cultural, and economic drivers. These change drivers can "push" against the boundaries of the system, compelling change to subsystems, and even forcing a redefinition of organizational purpose.

### 2. Values

The purpose of the organization implicitly reflects its values. Value statements may be a blend of existing characteristics and values to which the company strives. From the perspective of workplace design, a value statement such as "Employees' families are important to us" suggests facility design or policy possibilities such as internal day care, on-site dry cleaning, food service, telecommuting programs, and others.

### 3. Measurable Objectives

Business objectives will relate to the purpose of the organization, and may contain statements such as yearly production targets, number of new products developed in a given time period, or other strategic issues such as employee retention, attraction, and the like (see Figure 2.1). Objectives are measurable goals that are generally stable over time. In many cases it is possible to develop workplace design guidelines that support business objectives, directly or indirectly. For instance, design guidelines can be developed that create behavioral change (for instance, increased feelings of community or belonging to the company) that in turn support higher level business objectives, such as retention.

As another example, assume a business objective is to bring a certain number of new products to market in a given period of time. In order to accomplish this goal, collaboration between disparate teams and departments may need to significantly increase. Thus, workplace design guidelines could specify different design solutions that could be employed to enhance communication and collaboration. Like the business objectives, the success of these design objectives can be assessed through measurable outcomes such as a change in communication and collaboration through observations or self-report surveys. Measures in behavioral change can also be linked to the business outcomes themselves. Detailed discussion of our measurement model and approach, as well as case studies having these types of measures, can be found in Chapter 4 of this book.

## 4. Feedback System

The biological model has a built-in feedback system that connects information pertaining to quality of output with the objectives of the organization (see Figure 2.1). In a true biological system, such as at the level of groups (networks) of brain cells, a built-in feedback mechanism is required to maintain and optimize the behavior of the system (O'Neill, 1991).

In workplaces and organizations we discuss a related concept, that of "environmental control." Control is a mechanism by which information from the feedback loop is acted upon and used to quantitatively change the form and behavior of the physical and social subsystems.

Feedback and control are central elements of the biological model and are discussed in greater detail in later portions of this volume.

## 5. Scalability

The overall model is "scalable," that is, it can be applied to predict and understand behavior at different levels of the organization and workplace, including the individual/small group, team/business unit, or the entire organization. This scaleability permits us to apply concepts of the biological model to different scale design problems (individual workstation, group spaces, facility scale layout), and also permits the creation of measurement strategies to assess work effectiveness at those different levels of analysis.

## 6. Input

Input to the system can come in the form of data, ideas, or knowledge that form the building blocks of value-added products or services. Because our biological system is scalable the content of the input will vary depending upon the scale of the organization being modeled.

"Input" is shaped by the external environment, which is the political, cultural, and economic context within which an organization finds itself (Figure 2.1). The external environment may also act as a filter to block certain types of input from entering the system (to the benefit or detriment of that system). The input will vary depending upon the desired output. Thus, an R&D group process in which the outcome is a new product will have quite different inputs than that of a business unit that conducts consulting engagements. Thus, while "input" is a general term, it can be thought of as either a physical or intangible element that has the potential to be acted upon or transformed into something of greater value for a customer.

The input enters the social, technical, and environmental subsystems, in which some series of transformation events (business and work processes) act on that input (see Figure 2.1).

## 7. Output

In a well-designed organization, the output of the system (products or services offered by the company) should be consistent with the objectives (see Figure 2.1). In other words, a company having the goal of making great ice-cream will typically not attempt to offer computer software as a product. Because the focus of the biological system is on the output, we view the creation of workplace design as a means of facilitating the work activities and business processes that are required to create the output. This design may be at the level of managing adjacencies and block planning or decisions relating to consolidation of multiple locations, or it may be at the level of designing appropriate meeting and individual workspaces.

## 8. Throughput or Flow

Once we understand the purpose of the organization (or business unit, or department), we can move to understanding the transformation of "input" to the system into "output" (product). Figure 2.1 shows how input flows through the system, which includes the input, transformation, and output phases (note arrows through the system). As part of "throughput," multiple business processes cross departmental or group boundaries and are supported by the social, technical, and work environment subsystems.

## B. Boundaries

Boundaries are related to the limits of responsibilities of the organization (Taylor and Felton, 1993). Different organizations will have different boundaries. The overall boundary of an organization can be thought of as its "sphere of influence" in the marketplace, with its customers, employees, government, and competitors. There are four types of boundaries in our biological systems model, including: throughput, physical, social, and time.

### 1. Throughput

The throughput boundary starts at the point at which input enters the system and ends at the point where the output is delivered to the customer (see Figure 2.1). This is the defining boundary for the organization, since it involves the transformation of the input into the product or services offered by the organization (output). The quality of this throughput process must be closely aligned with the overall purpose of the system. Established roles and work responsibilities within the business processes supporting technology "throughput" are critical for success.

### 2. Physical

The physical boundary of the system is defined by the workplace occupied by the people doing the work. The biological system itself is anchored in the physical space. This space may occupy one floor within a large building, a campus of buildings, or a far flung network of corporate facilities, home offices, sales centers, and vendor and customer work locations around the world. Member obligations and responsibilities may go beyond the boundaries of a particular space or group of physical spaces. Given the distributed nature of knowledge work and the use of networked communications technology, the physical space may appear tangential to the work process. In other situations, the physical space in which work occurs may be controlled and occupied by other organizations.

However, far from minimizing the importance of the physical environment on business process, these trends in technology and work styles make understanding and effective use of office workspace even more important to success of organizations. Companies are beginning to understand the impact of workspace as a tool for communicating and enhancing corporate community, enhancing attraction and retention, and even for "branding" corporate identity to vendors, customers, and their own employees.

## 3. Social

This boundary is defined by the people directly involved in the work processes and the interaction between individuals and groups. Today, this "people" boundary is increasingly difficult to define since there are many classifications of workers, including: part time, freelance workers, workers on retainer, individual consultants, and external vendors that work to support the goals of the organization. To understand the social boundary of an organization, it is best to focus on understanding the roles that groups or individuals play in support of organizational goals, and not to use the existence of formal employment as a criterion for inclusion within the social boundary.

The social (or people) boundary is defined by the workers directly involved in the throughput of the system. In the case of knowledge work, the people involved would include not only technical and professional workers but their managers as well. The social boundaries of a manufacturing place would include production employees and their work group leaders. For all work, the social boundaries become extended and somewhat blurred with the inclusion of consultants and small service providers that work temporarily within the social boundary on a project basis. The social boundaries within the knowledge work systems grow and shrink along with the life cycle of projects existing within the current throughput of the system.

The dynamic nature of the social boundary has implications for the capabilities of the physical boundary in terms of accommodating frequent shifts in number of people at their work process, and supporting identification with the company and the role clarity of groups.

## 4. Time

The time boundary has to do with the time demands or constraints placed on the system in terms of producing timely output or product. The time boundary is greatly influenced by the purpose of the system. The time boundary of a system when the mission is to produce a seasonal product (snowmobiles) will be different from a system designed to exchange securities at a daily profit. The criteria for design of physical office space is influenced not only by the goals of the organization, but by time boundaries that influence effective work.

## II. CHANGE DRIVERS IN BIOLOGICAL SYSTEMS

Our discussion of the components of the Biological Systems model has thus far focused on the internal systems, processes, and goals of the organization. This model also incorporates the natural pull toward the future experienced by all living systems, and the external drivers of change that can cause radical shifts in the "rules" of a biological system overnight.

Our integration of the these change drivers is intended to make the model more robust in terms of understanding and predicting organizational and facilities change. A biological systems perspective allows us to consider external drivers of change that may affect the growth of a business, and ongoing business processes.

The notion of "environment" includes everything that lies outside the various boundaries of the system that we have described. An important goal of system design is to enhance the fit between the system and its environment, which includes the market and external stakeholders (such as customers, shareholders, suppliers, the local community, etc.). When the expectations between these stakeholders, and the activities of the system conflict, it may be time to re-examine the purpose of the system, its subsystems, or boundaries. When conflicts arise between parts of the system and some aspect of the environment, it may signal an opportunity to take the system in a new direction.

The role of change in our biological systems model exists on the "outside" of the organization, affecting the design of the system, which, as we have discussed, has dynamic but well-defined boundaries. At this point we consider what happens "outside" the boundaries of the system (see Figure 2.2). We will briefly discuss the types of conditions that are causal agents to change. Any one or more external conditions can serve to act as a "change agent" to the natural system. We discuss several business change drivers, including: globalization of markets, borderless finance and the migration of capital, and competition through growth, technology, and demographics (see Figure 2.2). These change drivers "push" against the boundaries of the system.

## A. Globalization of Markets

We live in an era in which modern capitalism has become globalized. The process of globalization consists of companies investing capital in foreign countries. This investment can take the form of buying existing assets, building new offices or manufacturing facilities, buying other companies, or other approaches. The business logic of commerce and capital has overcome established political boundaries and social orders, and is transforming nations (Greider, 1997). This economic revolution is fueled by invention and technology, and a desire to grow and accumulate wealth. Established rules of politics, respect of national borders, social protocols, and allegiance of country cannot stop the change toward a global market. The economic policies of home governments no longer play a singular role in business organizations' investment and trade decisions.

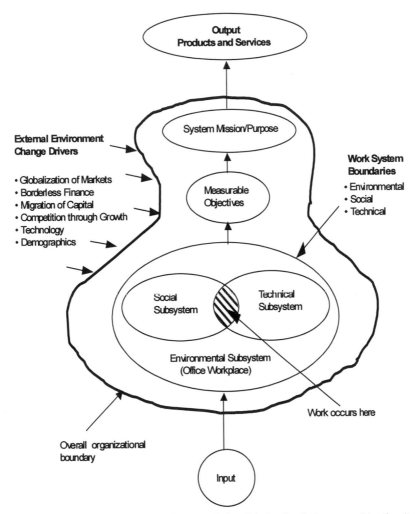

**Figure 2.2** External change drivers of the Biological System (Author)

Due to rapid advances in technology and improvements to worldwide service infrastructure, even relatively small U.S. companies now have access to world markets that were previously closed to them. However, these same small companies now find themselves in a market with competitors from their own and other nations who are grimly determined to win. Driven by competition, huge new business ventures are announced almost daily. U.S. companies are making sunglasses in India, and toothbrushes in Columbia. BMW picks South Carolina to build cars. Intel fabricates semiconductor chips in Malaysia. U.S. telecommunications companies form alliances with the national telephone companies of other countries. Boeing has agreements to build an ever-increasing proportion of its aircraft in other countries. While the largest U.S. companies have grown dramatically in sales, their worldwide employment has remained flat since the 1970s (Greider, 1997). The human labor required to create and deliver new services and products has consistently declined.

The traditional understanding of trade between nations is changing because of the trend of buying existing assets in other countries or building new factories so that products can be shipped for import back into the U.S. While U.S. companies have practiced this approach for years, other foreign companies have successfully adopted this approach within the U.S.

This steady dispersal of capital, production elements, and jobs across nations has the effect of distributing the "system" of the organization (illustrated in Figure 2.2) across a very wide spatial, geographic, and cultural space.

These trends create a tremendous strain on the internal system of the organizations experiencing these changes. The creation of new alliances may change the very purpose of the organization, creating a need for a redesign of the system and a rethinking of the design of the workspaces that span time zones and countries.

The distribution of professional and production workers across time zones and cultures brings challenges to systems design in terms of effective communication of corporate values, not to mention business goals. The environmental, social, and technical boundaries of the system may need to be redrawn in anticipation of the changing role of the organization (see Figure 2.2). Any changes to the system require time and may be painful in terms of relocated or eliminated jobs. There will undoubtedly be a time lag between when the time change has occurred and system redesign is accomplished.

## B.  Borderless Finance

The trading of stocks, bonds, loans, and other investment vehicles around the world continues to accelerate (Greider, 1997). Foreign exchange of currencies has increased even more rapidly, as traders move in and out of foreign currencies as they execute stock or bond trades. The entire global volume of publicly traded assets (about 24 trillion dollars U.S.) turns over about every 24 days. U.S. government bonds turn over even faster. The entire volume of U.S. bonds (about 2.6 trillion dollars) changes hands about every 8 days.

These enormous sums are traded by a relatively small group of banks and brokerages that trade on behalf of pension funds, mutual funds, and other banks and investor pools. This great amount of trading volume has increased the power of the global financial markets, and correspondingly the potential problem of market volatility across world markets and economies. News events or shifts in public mood can cause prices to change sharply. When investors lose confidence in one company or even industry sector, they can shift large amounts of money to another area almost instantly. This freedom of movement of capital has been facilitated by the virtual elimination of taxes on currency exchange transactions.

## C. Migration of Capital

All of the major industrial countries have experienced shocks when the global community moved investment capital. These major moves can influence the value of a country's currency and put pressure on its interest rates. National governments alone and in coordination with other nations have intervened in attempts to shore up their currencies when these shifts occur, but in many cases the strength of the global markets overcome even these efforts.

The cost of money, labor, and regulatory structure of countries influence where capital will be invested. Labor is particularly vulnerable to the migration of capital, since capital is free to choose from many labor markets at once, while labor is generally fixed in one location (Greider, 1997). While labor pools tend to migrate over time to geographic regions in which market conditions are favorable, the speed of movement of labor is no match for the instantaneous movement of capital.

The general volatility of capital and potential for rapid shifts in investment places intense pressure on business organizations and their subsystems. Fair or not, companies that do not respond to the demands placed on them by global investors may find themselves punished by the markets. Internal flexibility in organizations subsystems is key to anticipating the effects of negative market moves by the markets.

## D. Competition through Growth

Organizations react to competitors entering the market in a variety of ways. Increasingly, companies compete through incremental improvements to products -- and by depending on leaps in technology to improve development and manufacturing processes and thus reduce production time and costs. Thus, an increasing common reaction to competition is to improve an existing product, cheapen the cost to bring it to market, reduce its cost to the consumer, and flood the market with an oversupply of the product. The intention is to drive competitors out of the market through sheer availability of the product and low price. This, of course, also has the unfortunate side effect of commoditizing products and hastening the maturing of the market. Companies that cannot adapt to this challenge quickly enough, by improving processes and reducing costs, must withdraw from the market or be eliminated entirely.

When a decision is made to compete through growth, investments are made in process technology to improve product quality and reduce costs. Additional capacity must be pushed through the system. Existing facilities can prove to be a drag on process improvements brought on by technology and the increased volume demands of this growth strategy. It may prove more expensive to renovate a manufacturing or office facility than to start from scratch. Good examples of this problem are in the biotechnology and semiconductor industries. Facility planning and design must take into consideration the need for a great deal of flexibility in the internal layout and technical systems of a building, so that changes in technology, technical, and social subsystems can be rapidly accommodated.

## E. Technology

The new information technology allows workers to amplify their intellects, as opposed to the machines of the industrial revolution that leveraged muscle power. When computers were initially introduced into the office workplace, they were viewed from the same perspective as other machines -- that is-- in terms of their potential for leveraging the *quantity* of work that could be accomplished by a person or group; hence the popularity of early terms for computer applications like "word processing," or "data processing." This viewpoint followed the logic of the way other machines were used in mass production systems.

Over time and with increasing sophistication of hardware and software, the computer and its related technology have become tools to amplify intellect and, in themselves, create intellectual capital. Workers with high levels of technical skills are able to use this technology to leverage their knowledge and directly create economic value from their efforts. As has been pointed out in numerous books and magazine articles, the computer has eliminated an entire class of middle management jobs that were based on collecting information from one level of an organization, summarizing it, and passing it "up" to higher levels within the company. With wireless networks connecting people and information managers can now gain access to virtually any type of information they may need within their business. With this technology, operational decisions can be made that affect manufacturing facilities, and the activities of office workers, and even markets around the world.

This technology has given organizations a great deal of latitude in the way they are designed. For example, the "flattening" of hierarchy, dispersal of production capability, access to market channels, and so forth. The heaviest drag on the adoption of new technology is the existence of the status quo (Greider, 1997). The drag on change from existing factories, office buildings, social contracts around work, laws, unions, and politics is significant. When a new technology becomes available, whether embodied within a process, capability, or some other invention, the existing physical and social structure of organizations and communities stands in the way of change.

This resistance to change, or simple inertia against it, also lead to another significant driver of change -- the migration of capital. Investors may find it easier to move capital to new locations when the existing structure resists change. Smaller, poorer, less "developed" nations may lack the laws, social restraints, or even physical infrastructure that could act as a drag against new business practices that are driven by invention and technology. Desiring investment, these countries may be willing to accommodate arrangements that, if made in more advanced industrial countries, would require significant change to legal restraints, union agreements, and the like.

## F. Demographics

Companies face a dual challenge from the shift in demographics that is occurring in the U.S. This challenge is taking the form of addressing demographic change in both employees, and markets that companies serve. Gender and cultural diversity is increasing as people from a wide array of countries and cultures work for U.S. domestic and multinational concerns. This diversity in the work force is driving change in organizational cultural mores and expectations of employees that companies must address. For instance, Islam is a fast-growing religion in the U.S., and companies are accommodating the need of Muslim employees to pray several times during the traditional business day. In addition, cultural and racial diversity is also driving anthropocentric diversity in the work force, which means that work environment design must accommodate a work force with a wider range of body sizes than was previously the norm.

This demographic phenomenon is creating a set of changing expectations that exist within the markets that many companies serve. For instance, as a new middle class develops in countries such as India and China, demands for products and services will be influenced by the social and cultural milieu.

## III. THE BIOLOGICAL SYSTEMS MODEL OF GROWTH

Organizations pass through discrete phases of life, much like people or other living organisms, as they grow and change over time. As companies enter different phases of life, their needs in terms of office work environments and overall facility strategy will change as well. To better understand the processes and phases of change, we draw upon Land and Jarman's (1993) natural systems model. Thus, we view our biological systems model as a powerful descriptor of the internal components mechanisms and processes that drive the organization and workplace.

We employ Land and Jarman's natural systems model as a means of predicting and understanding the *phases of growth and change* within our biological model. This section provides an overview of that three-phase model, which includes stages of forming, norming, and integration.

The natural systems model is a means of understanding the role that external change forces play in organizational growth and change over time. Thus, the natural systems model allows us to understand the changes in form that the natural system must take as it reacts to change drivers and grows and changes internally over time.

### A.  Breakpoint

The key to understanding this model is the notion of "breakpoint," or rapid, qualitative change that is occurring in organizations and in society at large. This type of change is different from the type of change we have normally experienced in the social and business arenas, at least in the U.S. Breakpoint change is a change in the underlying "rules of the game" within a business or other organization.

For instance, at the turn of the century the railroad industry contained some of the largest, fastest growing, and wealthiest companies in the U.S. In the early decades of this century the fledgling airline industry was born, at first carrying only mail and packages and then passengers. The early airline companies were weak and vulnerable, but they represented the future of rapid travel for business and pleasure. The railroad companies could have easily bought the airline companies and controlled that new part of the travel industry as well. However, the railroads ignored this new transportation mode as a passing fad because they looked to the rules of travel that had applied in the past, instead of realizing that the rules of passenger travel essentially had changed overnight. It took only a decade or two for the railroad passenger industry to collapse in the U.S.

Another example of breakpoint change is with the movie industry and the fledgling television industry in the early 1950s. The movie industry was wealthy and powerful and simply could not recognize the powerful shift in preferences and viewing habits that would occur with the introduction of television programming. Who would have thought that people would rather sit at home and watch a box, than go out to a movie? This rapid shift in preferences and life patterns of consumers took the movie industry completely by surprise. A breakpoint shift in the rules governing entertainment consumers had taken place and practically destroyed the movie industry. The people providing the media content saw a shift in the rules of stardom as well. In the movies, an actor could become a star and enjoy a certain stability in their fame over time. Because of the much faster pace of the television media, actors became stars much more quickly, and faded more quickly through a mixture of overexposure and the media's voracious appetite for new people and content. Thus, another example of breakpoint in the fundamental rules of stardom as we know it today.

Examples of breakpoint change abound in society, within political institutions, within industries, and in individual organizations. Breakpoint change cannot be predicted by an understanding of past events. Traditional change is quantitative in nature, meaning that if we had x, y, and z in the past, we will predict the future by saying we will have *MORE* x, y, and z, and we'll have it faster. Predicting the future by relying too heavily on the past can result in decisions that are snapshots of the past, reacting to past rules and events.

Business decisions and judgments about facilities and work environments run the risk of becoming reactive to past events, instead of anticipative of future states. The risk increases if the organization is about to enter, or is already within a breakpoint phase. To understand more about how and why breakpoints occur within organizations, it is necessary to understand the phases of organizational growth and change that accompany breakpoint change (Land and Jarman, 1993).

This natural systems growth model is illustrated in Figure 2.3. The model suggests that organizations, in general, follow a three-phase cycle of growth over time. These phases are known as "forming," (Phase 1), "creating norms," (Phase 2), and "integrating" (Phase 3).

## B. Forming

The first phase of growth is called "forming." An organization in this phase is usually entrepreneurial in nature. Figure 2.3 illustrates the forming phase. Start-up companies are an example of organizations that might be in this phase. The culture of a company in this phase is fairly fluid, probably influenced by the founder and/or owners of the company.

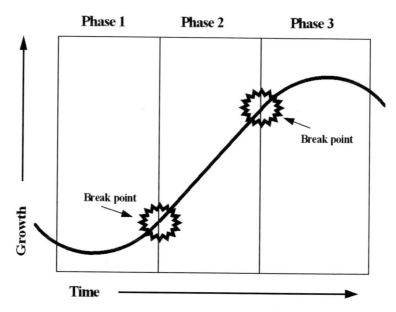

**Figure 2.3** Phases of organizational change (Author)

The thinking is divergent, many ideas are being considered (and rejected), there may be lots of creative energy on the staff, and every effort is "trial and error" as the organization tries to find out who it is, what the product will be, how they will go to market, and everything is open to discussion. The physical work environment is probably not of great concern to the workers at this point. In fact, workers may pride themselves on "overriding" the design intent of existing work environment configuration and layout by moving furniture around, bringing in items of their own (refrigerators, couches, etc.), and using the existing environment in unconventional ways.

Then at a certain point of growth in this phase, ideas and mission become focused, and the company enters a change breakpoint (see Figure 2.4). At this point, the rules underlying the desired behaviors and values may rapidly shift. As the company enters the market with their first (or new) product, the divergence and creativity that was once valued is now seen as undesirable, since it might diminish the focus on the task at hand, which is to do the thing the company has finally decided to do (new product, service, or customer). Inventive activities and the trial-and-error approach are seen as wasteful of resources and time, when there is a job to do. If the people within the organization do not understand that a breakpoint has occurred, they might not engage in activities that will contribute to the immediate needs and success of the effort. The company has now entered Phase 2 of growth, the "norming" phase (see Figure 2.5).

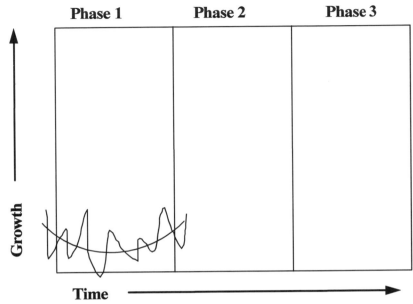

**Figure 2.4** "Forming" phase of organizational growth (Author)

## C. Norming

In the second phase of growth, the growth curve for the company is steep and relatively consistent (see Figure 2.5). More workers are added and there may be a transition to professional management if the company was initially run by its founder. In this phase, the emphasis is on product extensions and incremental improvements, rather than on the introduction of completely new products or ideas. The company begins to attract employees with skills in operations, and the focus of management shifts to measurement of internal processes as opposed, perhaps, to customer needs. The company is doing well financially, and the culture is one of "don't rock the boat."

Professional management of the physical work environment may become a necessity, and the need to suddenly house a rapidly growing number of people with some degree of economy and organization may result in the adoption of a mechanistic orientation toward the design and provision of work environments. As the organization enters deeper into Phase 2, the mechanistic orientation toward the work environment becomes the norm, with only minor variations in work environments permitted (such as square footage or degree of enclosure).

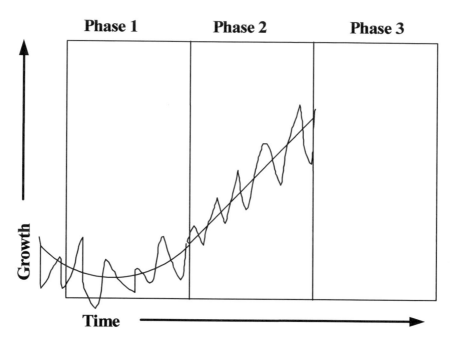

**Figure 2.5** Phase 2 of organizational growth (Author)

Any allowable variation quickly becomes seized upon as a means of broadcasting status or other information about individuals. Companies may institute policies that reinforce the mechanistic attitude toward variance (which is to say an intolerance towards variance), such as "clean desk" rules in which worksurfaces must be cleared at the end of the work day. In a sense, the work environment is aligned with the organization. Often, organizations in Phase 2 are like finely tuned machines with many parts. The parts mesh together tightly and are designed for efficiency. The environment is part of that efficient machine. The machine works well until it needs to change (due to market forces, etc.). Then the very efficiencies and tight variances that made the machine so successful cause it to fail in the face of a breakpoint and rapid change. At that point the environment can become significantly misaligned.

According to the biological systems model, it is impossible for a natural system (whether a society, a person, or an organization) to sustain that type of growth indefinitely. The company at this point has been successful for some time, and is well established in its market. Suddenly, competitors appear with lower prices for functionally equivalent products. Profit margins are squeezed. Perhaps some fundamental need of the market has shifted without the realization of management. Management responds with a renewed emphasis on controlling internal costs and streamlining processes. This emphasis on costs may unfortunately come at the expense of head count within the company. This renewed focus on operations and internal costs may temporarily boost growth and profitability, but the organization has reached a breakpoint. Land and Jarman (1993) call this the "back to basics" bump (see Figure 2.6)

The behaviors that for so long were appropriate for a Phase 2 company, now only delay the day of reckoning. If the organization survives the Phase 2 breakpoint, it enters into Phase 3, which is called "Integration."

## D. Integration

In this phase, growth remains relatively flat, at least compared to the heady Phase 2 days (see Figure 2.6). The organization begins to re-examine ideas that were developed in Phase 1 and long ago rejected by the company while it was in Phase 2.

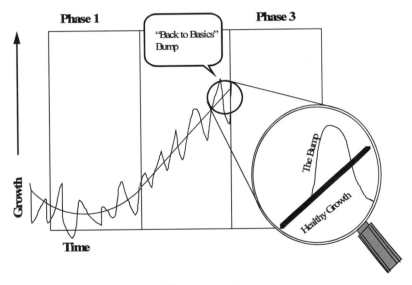

**Figure 2.6** "Back to Basics" bump (Author)

One or more of these ideas may be integrated into the Phase 3 thinking of the company. This is a difficult time because the new ideas may demand funding at a time when resources within the company are tight. This remains a phase of marginal growth and fear of the future, and a longing for the "good old days" of Phase 2.

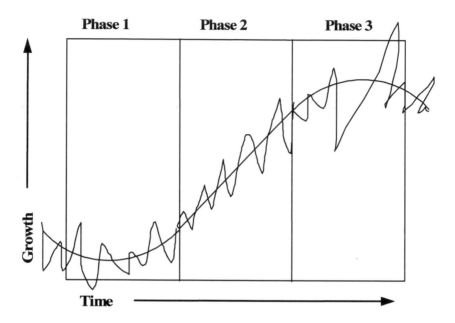

**Figure 2.7** Phase 3 of organizational change (Author)

This is also a phase of corporate reorganizations and continued downsizing, especially if the organization remains committed to a future that is based on its past. Efforts to integrate new ideas may meet with initial enthusiasm followed by disappointment when rapid financial growth fails to follow. In order for a company to truly thrive after this phase, it must reinvent itself and recreate the first phase patterns.The growing misalignment of the physical work environment with this attempt at new directions becomes obvious to workers and management alike, although few can articulate their feelings.

## E. Re-invention

In order for a company to truly thrive after this phase, it must reinvent itself and recreate the first phase patterns. We call this phase "reinvention," because it is at this critical juncture that an organization must select ideas that were once considered in the original Phase 1, and free up the resources to pursue those ideas. Figure 2.8 illustrates a model of reinvention, showing a new Phase 1 being created at the same time the organization continues through Phase 3. This is a dangerous time for companies, because just at a time when internal resources (money and time) are at a premium, the company must take resources and apply them to new, uncertain ventures. The risk runs high that after a short period of time the parent organization will interfere with the somewhat less systematized, less proceduralized activities of the reinvention phase. The more "freewheeling" culture of any new venture will quite possibly clash with the culture of Phase 2 or Phase 3 organizations.

The physical work environment can help to play a role in supporting reinvention within an organization, and it may also act as a catalyst for change. The environment needs to be designed from a biological perspective. If the environment in which reinvention is occurring can give the new venture secure yet dynamic boundaries in which to form and thrive, the process of reinvention stands a better chance of success. The boundaries of the facility must also be permeable so that appropriate interaction between people involved in reinvention activities, and people involved in running the existing core business, can easily interact. The flow of people, ideas, and resources across this boundary will change over time and with the needs of both areas of the company. The physical space needs to not only provide boundary management for both "sides" of the company, but must provide support for a unique culture identity for the support of invention activities throughout the organization.

## F. The Integrated Biological Systems Model

Figure 2.9 illustrates the integration of the concept of dynamic change into the model. As described in earlier sections of this chapter, the internal system of the organization is comprised of social, technical, and work environment subsystems. In a static environment, it is possible to align these subsystems to achieve the desired objectives and ultimately the mission of the business.

However, our Biological Systems model integrates several dynamic elements that constantly seek to move the internal subsystems out of alignment with the objectives of the organization. The first of these elements is the external environment. The external environment consists of forces such as the behavior of capital, technology, demographics, and other extrinsic drivers that constantly push against the system and cause misalignment.

The second dynamic force is the natural, internal growth of the system as it changes and moves through various phases of qualitative development. Figure 2.9 illustrates the dynamic pressure of the internal subsystems as they push from the inside against the form and boundaries of the business, changing the form and boundaries of the organization over time. Figure 2.9 shows the effects of change on the environmental subsystem. While the rest of the components of the organization have moved, the environmental subsystem has remained in its original position relative to these other systems.

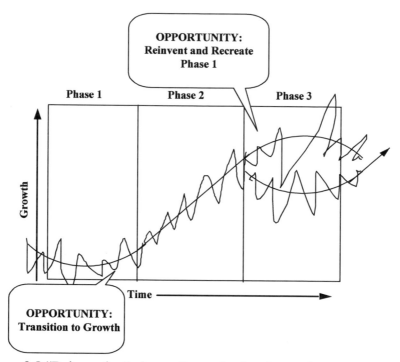

**Figure 2.8** "Reinvention" phase of organizational growth (Author)

Thus, while the rest of the organization has changed, the physical work-space has lagged behind and is not properly supporting the social and technical systems, nor is it in alignment with the objectives and purpose of the larger system.

For many companies this is an opportunity to re-align their physical work environments with both their changed internal structure and work systems. Control over the physical environment, whether exerted at the individual, team, or organizational levels, is the means by which this alignment can be achieved. This concept of control is considered in more detail in later sections of this volume. Control must be paired with purposeful analysis and design to be effective.

## IV. FACILITY DESIGN FOR ADAPTIVE CHANGE

In our practice, we approach projects by thinking about these issues at different levels. Figure 2.10 shows our approach to problem solving, and understanding and predicting change.

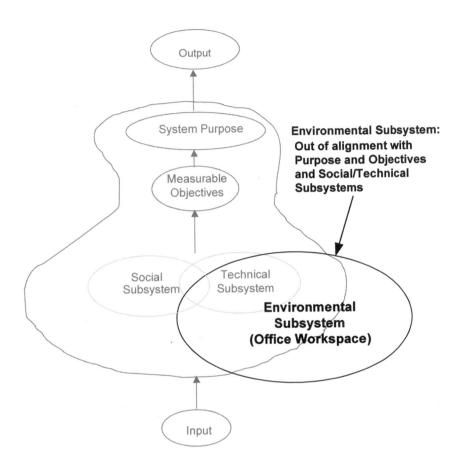

**Figure 2.9** The Integrated Biological Systems model (Author)

Unfortunately, problems can occur when work environment design issues are addressed in isolation from other systems within the organization, such as the technology and social systems. For instance, when a group or department moves, it provides an opportunity not only for redesign of the space, but for developing design solutions in the broader context of the needs of the organization. Such pilot projects can be a positive means of testing new ideas and integrating business process and technology solutions.

In our problem solving approach, we consider the broader context in which the organization exists (see left, Figure 2.10). As part of the context, we consider change drivers, culture, the business mission, and current phase of growth for that company. Is the company approaching a breakpoint? We develop specific metrics around the workplace to measure its current performance related to business objectives, and to assess performance after the redesign of new space.

| Context | Analysis | Alignment |
|---|---|---|
| ☐ Change Drivers | ☐ Workplace | ☐ Workplace Design |
| ☐ Culture/Values | ☐ Technology | ☐ Technology |
| ☐ Business Objectives and Metrics | ☐ Business Process | Requirements |
| ☐ Phase of Growth | | ☐ Business Process |

**Figure 2.10** Problem solving approach (Author)

Once we understand the context of the problem and appropriate performance metrics, we can analyze the subsystems within the organization (workplace, technology and business process), to determine why it is structured the way it is, how individuals and groups interact, and how organizational success and work effectiveness are defined (see middle of Figure 2.10). This analysis supports the alignment of workplace design and strategy (see right, Figure 2.9) with the needs of the organization. Thus, it is quite possible that a work environment solution could take a significantly different form given this contextual approach than if the solution is approached in isolation. We propose a cyclical approach to maintaining alignment of the physical workspace with the organization, and work effectiveness (see Figure 2.11).

As we have discussed to this point, work effectiveness must be approached by making an inquiry into the existing structure and activities within an organization. In order that the solution not be a snapshot of the past, the development of design criteria must use a "future orientation," in which a future state of the work environment (including technology and business processes) are designed. At this point, metrics for assessing the performance of the new work environment should be established. These metrics should be closely related to the stated business goals of the organization, business unit, or group for which the design is being developed. A participatory process for design development should be used that involves end-users and their representatives in both gathering required information, and evaluating design concepts as they are developed. Once the new space is implemented and people have moved in, data should be collected to assess the success of the new space against the performance criteria developed at the start of the project. Because of the dynamic nature of organizations and the constant change they experience due to internal growth and external change drivers, we see this change process as cyclical in nature.

Participatory
Design and
Implementation

Evaluation
and
Feedback
into
Next Phase

Criteria Development
Future Orientation
Establish Performance Metrics

**Figure 2.11** Cyclical approach to design of effective office work spaces (Author)

Finally, and perhaps most importantly, we believe that these processes must result in *actionable change* that can be clearly implemented in some form. The objectives of the change to the work environment must be clearly articulated so that the success of the change program can be evaluated. Thus, we argue that the value of any intervention must be measured by the yardstick of objective measures of individual, team, or organizational performance. Other measures may include facilities asset use (space efficiency), or simple return on investment (ROI) models. For clients with more sophisticated demands, we have developed financial modeling tools that illustrate the effects of work environment improvements on the leveraging of compensation for entire company divisions. These measures may be obtained through various means, but should always be developed in conjunction with the company managers so that the results are internally valid to the organization.

## V. DESIGN ASSUMPTIONS OF THE BIOLOGICAL MODEL

The tools used within the Biological Metaphor do not focus on "workstation design" issues. The outcomes are strategic design guidelines that support organizational objectives. Unless couched within strategic design guidelines, typical workplace design approaches will do little to enhance organizational effectiveness. This approach goes beyond the individual or group level design issues, to address problems at the level of the business unit or entire organization. How large a piece of the organization is defined as a system in this approach will vary from case to case. Thus it is possible to design an entire organization using this approach, but in practice it is more typically used to design spaces for departments or business units.

## A. Assumptions

This approach contains a number of important assumptions:

Business organizations can be thought of as biological systems, containing subsystems made up of people, technology, and the physical environment. These systems change over time in response to external events and internal growth and learning.

The focus of the Biological systems approach is on the goals or output of the business organization, rather than on individual tasks or processes. The product of the system should be closely related to the purpose and objectives of the organization.

Business organizations, and their work environments, can be "purposefully designed" to achieve organizational goals.

Ours is not a static model. Change resulting from drivers in the external business environment continually "push" against the boundaries of the system and require internal change to the organization and the environment. Change can be anticipated and built into the system.

We value communicating change before moving people into new spaces, especially if those spaces are very different in concept to what they are accustomed to. Employees can understand changes to the work environment if they have shared understanding for the reasons behind the change.

When using this approach to create new workplace designs, a shared understanding is created within the organization, focusing everyone on the same goals for design. It is first and foremost a participatory process. In this approach, decision-making is pushed "down" throughout the organization. In this way, employees "own" the change (or at least feel like they do). This also includes participation in collecting information pertaining to the design of the office work space and being asked to provide feedback on design schemes throughout the process.

The Biological Systems methodology is not a standardized set of steps that are followed without variance for every case. This approach is more in the form of guidelines that should be followed. These guidelines are flexible enough to be modified depending upon the system that is being redesigned.

The purpose of the organizational and physical design must be known to all affected employees. Representation and participation in the process by at least a sample of affected members within the system is required, if the design process is to succeed.

Our methods are not intended to be used as a means of identifying problems within a system. All organizations and work environments have problems.

Rather, our goal of "purposeful design" keeps the focus on aligning the elements of the system in support of the goals of the organization. The Biological Systems approach is about system design, not system repair.

As part of the system design, metrics for success should be identified and agreed on at the start of the project. These metrics should be related to business objectives. The metrics can be used not only to guide design decisions but can be used to evaluate success of the project upon its completion. In Chapter 4 we discuss a detailed Measurement Model that integrates with the design development approach we discuss in this chapter, as well as the biological metaphor.

## VI. CONCLUSIONS

The dynamic force of the system as it grows internally, and transacts with the various extrinsic forces, also causes misalignment between subsystems. This internal growth dynamic is particularly damaging to the alignment of the environmental subsystem, which is prone to be ignored for long periods because of the costs associated with work environment change, and the lack of understanding of the contribution of the physical work environment as a lever for organizational effectiveness.

This chapter makes clear that any attempt to align workplace design with organizational goals presents significant challenges in understanding and aligning what are essentially a set of moving targets. Recent research (Light, 2005) supports our Biological Systems model and the values we have discussed to this point in the book (such as, future orientation, participatory design, job control, adaptability, aligning work systems and workspace with the mission of the organization) associated with creating effective work environments. This research has identified relevant characteristics of successful organizations, such as: "agility" (empowered employees and participatory management), "adaptability" (changing with circumstances to take advantage of new opportunities), and "alignment" (aligning the organization around the mission). While Light's model emphasizes the social and technology subsystems characteristics, these general characteristics are also components of our Biological Systems model -- which extends these ideas to the physical workspace as well.

The dynamic nature of the biological system, coupled with the relatively static nature of the typical office work environment suggest that the processes used to develop effective work environments need to be future-oriented and anticipatory in nature.

It also suggests that significant "proof" about the linkage between the physical space in which work occurs, and organizational effectiveness, needs to be made in order to justify the shift in thinking about the relevance of the work environment to organizational goals. The following chapter in this volume explores an alternative way of thinking about work environments, so that this "workplace effectiveness" linkage can be made at different levels within the organization. In later chapters, we suggest a framework for participatory and future oriented processes that can generate the level of understanding necessary to attain the strategic alignment of work environments with business mission and direction.

# CHAPTER 3

## Environmental Control and the Function of the Biological System

"We shape our buildings and afterwards our buildings shape us."
Winston Churchill (May 10, 1941)

The focus of this chapter is on the concept of "environmental control." Environmental control is the degree to which individuals, groups or business units can modify or adapt features of their physical workplace to enhance work or business processes towards achievement of business goals. We will explore two key notions about the role of "environmental control" in effective organizations.

The first idea is that environmental control is a primary mechanism within the biological metaphor that permits "self-managed" change and adaptation within organizations and their workplaces. Thus, while the workspace does shape our behavior to some extent, in a biological system, people in turn have the ability to shape their environment.

The second idea is that environmental control supports "job control." Job control has definitively been shown to enhance performance and health in individual workers. In our model we expand the concept of job control (and environmental control) beyond the individual, to incorporate groups and business units or departments.

## I. INTRODUCTION

In order to successfully compete in today's global markets, organizations are decentralizing decision-making authority, and in some cases reducing the role that middle management plays (DeGraff and Lawrence, 2002). This is made possible, in part, through the widespread implementation of effective information technology within organizations (Light, 2005). As a result of these trends, job definitions are broader and more complex. Individuals have greater workloads and responsibilities, and team-based activity has become the standard mode of getting work done. Employees and teams are increasingly expected to act autonomously in carrying out their work goals and team mission (O'Leary-Kelly et al., 1994).

A substantial amount of research shows that decision latitude, control over pace of work, location of work, and other related "job control" characteristics are related to stress, health, learning and performance (Frese, 1989; Karasek and Theorell, 1990, O'Neill and Evans, 2000). Jobs with high job control and high demands have been linked to positive health and performance outcomes. In jobs with high control, demands are seen as challenges. In high control jobs, stress arising from demands is a positive characteristic because high control jobs permit active response and full utilization of skills. These studies suggest that jobs can be "redesigned" to feature high decision latitude and participative decision making. To this end, many organizations provide training and development programs to encourage autonomy and flexible, creative behavior in their employees.

In our biological systems model, we suggest that this concept of job control can be extended beyond individuals, to the design and structure of work teams and even entire business units.

We also introduce a concept central to our model, that of "environmental control." Environmental control is the degree to which the individual, group or business unit can modify or adapt various features of the workplace to support work or business processes or other required changes to the system.

*Control (both environmental and job control) serves as the key mechanism by which the biological system (the business organization) can create and self-manage change to its internal structure and operation.* Control permits the organization to respond to changes driven by internal growth and by external demands and opportunities. Thus a central argument of this book is that organizations should explicitly design facilities and work spaces to provide and enhance environmental control.

Environmental control can be enhanced through the adaptability and flexibility of the workspace, and workplace design features that offer control will differ depending on the scale of the organizational unit. Thus, features that enhance environmental control for individual work are different than workplace features that enhance it at the group level, and are different still at the scale of the business unit.

Unfortunately for many companies, the office workplace still reflects the "machine metaphor" (see Chapter 1) and not the "biological metaphor" for workplace design. By its nature, the biological metaphor leads to adaptable and flexible workspace features. This chapter discusses the model of environmental control, how it relates to job control, and how it acts as a key adaptive mechanism within the biological model. In later portions of the book we discuss numerous studies, conducted by the author and others, that show the impact of workplace design for environmental control and its effects on job control and performance.

## I. ENVIRONMENTAL CONTROL

In this book, we argue that the physical work environment is itself a "lever" that the organization can manipulate to enhance control and ultimately, effectiveness. In this model, control can be exerted at different scales of the workplace, ranging from the facility scale (control at the Business Unit level) to the individual workstation (individual job control).

Thus, the opportunity for organizational effectiveness lies in the use of the facility as a mechanism to provide environmental control, through the application of technological and physical flexibility features that support changing work styles, organizational models, and business requirements.

Environmental control can be "designed in" at different scales of the organization. At the individual level, work station features (such as seating, task lighting, storage, shelving, work surface height, enclosure, VDT and keyboard, HVAC) can be designed to support work flow through user-adjustability and flexibility.

At the work group level, overall layout of the work environment can be designed to support the group's ability to self-manage and reconfigure boundaries between themselves and other parts of the organization, depending on business requirements.

At the organizational scale, the facility can be planned to enhance flexibility of interior building layout (reusability, integration of technology, ability to expand or downsize through physical reconfiguration).

## A. Control and the Biological Systems Model

In this volume, we discuss how control over the physical environment influences the effectiveness of work. Poor work performance, stress, and health problems may be related to lack of "fit" between individual job demands and the physical work environment. The physical work environment provides a mechanism of control to optimize the environment to match the demands of the job. Other research has examined the degree of fit between work demands and the organizational environment, and resultant stress (Edwards and Cooper, 1990). The primary intervention approach is to improve the adaptability of the person to the work environment. There is, unfortunately, less organizational concern with adapting the job design or physical environment to fit the worker (Cooper and Cartwright, 1994). Environmental control may provide a mechanism for workers to actively adapt the environment to fit work demands, rather than the other way around.

Our Biological Systems model uses a systems viewpoint to understand the effects of the environment, and control over the environment, on worker health (Altman and Rogoff, 1985; Smith and Sainfort, 1989). As discussed in Chapter 2, our approach incorporates the environmental, social, and technical subsystems that comprise the business organization.

Within this framework, the biological model is scalable; that is, it can be applied to different scales of the organization and supporting work environment. The "environmental subsystem" can be the individual workspace, the space supporting a business unit or department (which can be an area within the building), or an entire facility or campus (see Figure 3.1). Figure 3.1 represents the different scales of environmental subsystems involved for an individual employee working within a workstation, a team or department within a collaborative space, and all the business units within a facility. For each level of the environment, the scale and specifics of the social and technical subsystems will differ, but the fundamental elements of the system remain.

The model shows that, regardless of the scale of the environment being examined, there is still input to the system, a transformation event (work), and output that should be aligned with the purpose of the organization, business unit, or individual. Along with the scale of the environment, this model also recognizes that the scope of "Purpose" will vary along with the scale of the environment and group size under consideration (Figure 3.1). For instance, at the facility scale (scale of the organization), the scope of purpose would be the purpose of the company. At the departmental/business unit level, the purpose of the system is business unit mission or chart of work (see Figure 3.1). At the individual level, individual purpose (individual performance plan) may be considered.

*In each case, our model suggests that it might be possible to align the design of the work environment to the scope of the purpose appropriate to the individual, group, or work unit. Further, we contend that environmental control is a primary, self-managed mechanism for aligning the physical environment with the purpose of the system.*

Regardless of the scale of the environment being considered, the model specifies that the person-environment elements interact with the other subsystems in a reciprocal, modifying manner, through the mechanism of control. Thus in our model, the act of exerting control over the environment (and thus modifying it) can result in subsequent changes to the behavior of other components of the system.

The biological model emphasizes the dynamic nature of the processes underlying change within the system, and at each level or scale of the person-environment subsystem. In this model, control is seen as an overlying axiom that can be exerted at different levels within the system through a variety of mechanisms.

Because control is conceptualized as a means of accomplishing goals, the behavior of this system is seen as being "pulled" towards a state of ideal function -- that is, the effective accomplishment of organizational mission, and group or individual work goals.

## B. Conceptual Model of Environmental Control

Effective work is measured by the success with which individuals, groups, and organizations attain their goals. Job characteristics such as decision latitude (job control) and ongoing learning are related to effectiveness (Karasek and Theorell, 1990).

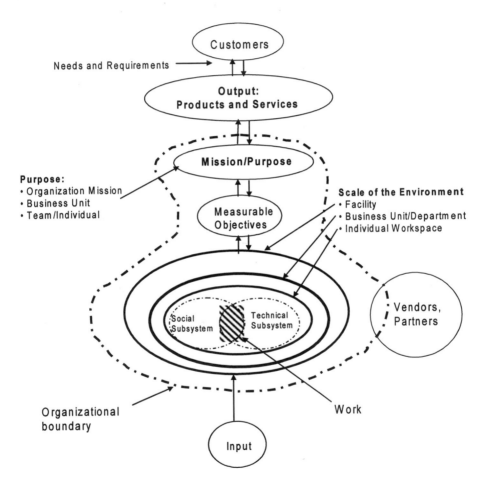

**Figure 3.1** Biological model showing different scales of the environment (Author)

Effectiveness may be reduced because of job stress, which can reduce the capacity for learning over time.

The model proposed in this chapter suggests a positive relationship between control over the physical work environment and work effectiveness (see Figure 3.2). Environmental control is defined as the degree to which the organization, group, or individual can exert control over the physical environment as part of the process of achieving work goals.

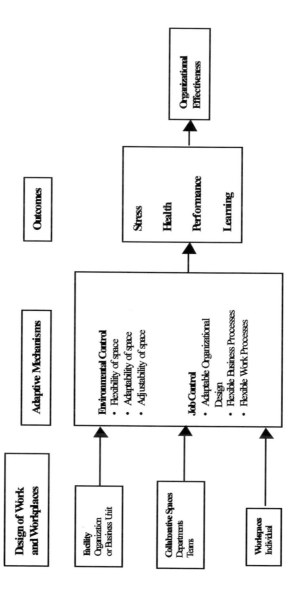

**Figure 3.2** Conceptual Model of Environmental Control (Author)

While the primary focus of this discussion is on effective work and environmental control, other constructs such as job control, learning, and stress are considered in the model (see Figure 3.2). The model recognizes the influence of both the psychosocial and physical work environments on effective work. Certain characteristics of each of these environments can act as risk factors to worker health. Within the psychosocial environment, a significant body of research indicates that characteristics of the job, such as psychological "demands," interact with the degree of job control to influence stress, learning, and performance (c.f. Karasek and Theorell, 1990) (see Figure 3.2.).

This model extends the concept of control over the job - to the physical environment. The design of the physical work environment is thought to provide opportunities for environmental control that might, in turn, affect job control, which is linked to stress, health, learning, and performance, and ultimately to organizational effectiveness.

## C. Review of Model Components

Beginning with a review of the concept of organizational effectiveness, this chapter will discuss each of the components of the model in turn, and then focus in detail on the mechanisms of environmental control and the different organizational and environmental scales into which opportunities for control can be designed.

## 1. Effective Work versus Productivity

It is evident that traditional notions of measuring worker productivity need to be re-thought. In this chapter, the traditional notion of productivity is contrasted against more recent thinking on worker "effectiveness.

Shortly after the turn of the century, Frederick Taylor (1911) developed an approach to designing organizations and jobs using "scientific" methods that are still practiced by many companies today. These methods are closely aligned in concept, with the machine metaphor we discussed earlier in this volume. Using this approach, workers' jobs are redesigned into a series of simple tasks. While this approach was initially developed to increase productivity on manufacturing jobs, it has been extended to the white collar workplace (e.g., data entry clerks, forms processing, telecommunications operators, and others). The emphasis of this approach is on achieving task efficiency through the optimization of work processes. The focus of productivity measures is on outcomes; that is, the amount of work performed in a given unit of time.

Taylor's approach cannot be applied easily to the design or analysis of knowledge work. Professional workers in today's information-rich work environments operate in a setting in which goals need to be met through group decision making, individual responsibility for job skills improvement and learning, and a technology-supported, autonomous work style. For these workers and teams, performance is measured by the effectiveness by which they attain their goals. Thus, effective work is measured as a function of mission and goal attainment, rather than through units of production (O'Leary-Kelly et al., 1994).

The concept of effective work has also been applied to teams (Sundstrom and Altman, 1989). Team effectiveness has two components: process and goal attainment (Jones et al., 1983; Nadler, 1977). The process component includes the quality of group member interaction, group stability, communication and problem solving (Goodman, 1986). The quality of the team process is also thought to influence the "viability" of the team in terms of its future prospects (Sundstrom, De Meuse, and Futrell, 1990). Goal attainment is measured through the acceptability of the team's output to customers within or outside the organization.

## 2. Job Control

As organizations continue to decentralize decision-making authority, and change the role of middle management staff, we have seen a corresponding increase in the desire to leverage the workspace and technology to enhance the mobility and flexibility of individuals and teams. As a result of these trends, job definitions are becoming more fluid and complex. Individuals have greater workloads, responsibilities, and reporting relationships, and team-based work is the norm. Workers and teams are increasingly expected to act autonomously in carrying out their work goals and team mission (O'Leary-Kelly et al., 1994). A substantial amount of research shows that decision latitude, job skill utilization, and other related job control characteristics are related to reduced stress, and enhanced health and learning((Frese, 1989; Karasek and Theorell, 1990). These studies argue that jobs can be "redesigned" to have high decision latitude and participative decision making. In jobs with high control, demands are seen as challenges and learning opportunities. The stress arising from demands is a positive characteristic because high control jobs permit active response and full utilization of skills.

## 3. Learning and Stress

The increase in individual decision latitude in jobs has increased personal responsibility for good decision making. Ongoing learning is important to effective work, because skills development can increase the quality of decision making. Karasek and Theorell (1990) suggest that there is a symmetrical relationship between stress and learning. They suggest that the cumulative effects of occupational stress over time may lower learning capacity (and thus worker effectiveness), because job demands prevent the worker from attaining a cognitive "rest state" between work periods. They also suggest that job-induced learning may reduce stress through the development of individual potential and self-esteem. Increasing job skills permits workers to meet job challenges with less stress. Numerous studies show that skills training and mastery over job content is consistently related to reduced psychological stress levels and physiological stress responses (Bruning and Frew, 1987). Stress has costs to the organization beyond reduced learning and effectiveness. The total cost of stress to U.S. business through absenteeism, reduced effectiveness, and insurance and compensation claims is over 150 billion dollars per year (Karasek and Theorell, 1990). Because of increased job autonomy, learning and ongoing job skill enhancement is critical to good decision making and work effectiveness. Many organizations provide training programs to encourage autonomy and learning skills in their employees (Land and Jarman, 1993). These links between learning and stress suggest that effective work must be considered within the context of a learning organization.

## 4. Risk Factors of Stress

We argue that there are two significant sources of stress in the work environment: 1) the way jobs are designed, and 2) the design of work environments. Jobs, like work environments, are "designed." Thus, they have certain characteristics that enhance or detract from the quality of work life.

*Job Design and Stress.* This section begins with a discussion of the way jobs are designed and how job design is related to stress. Jobs can be thought of as having two characteristics: psychological demands and decision latitude (Karasek and Theorell, 1990). Examples of the psychological demands of work -- essentially, "how hard you work" include deadlines, how many widgets you make per hour, how many reports are due this week. Demand is related to the need to maintain high levels of concentration for long periods of time -- the mental intensity of the work. However, just because the worker has a psychologically demanding job doesn't mean that he will experience stress.

In this model, the amount of stress experienced is dependent on the amount of decision latitude, or "control" available over the job. Decision latitude is the degree to which workers can employ their natural skills and talents on the job, and the freedom to make decisions about how to do the job. It also has to do with the amount of control available over the pace or rate at which tasks must be completed.

Karasek and Theorell (1990) showed how these two job characteristics (demand and control) can be plotted against each other to predict how stressful a particular job type will be (see Figure 3.3).

*Active jobs.* Active jobs reside in the upper right quadrant. Active jobs are those in which there are high demands, but also a high amount of control over how to do the work and freedom to be creative in work. This sort of professional work requires high levels of performance but workers also possess a high level of control and the freedom to use all their skills to perform the tasks associated with these jobs. An active job has many stressors because of the challenges inherent to the job. The stress resulting from this job type is translated into action as people actively solve problems, so very little of the stress is carried over to cause psychological or physiological problems. This not to say that active jobs are relaxing -- however, people can channel the energy into learning and problem solving, rather than allowing the stress of the job to affect them negatively.

*Low Strain Jobs.* Low strain jobs are jobs with few psychological demands and high levels of control. An example of a job type like this is an appliance repair person. These people have low risk of stress and illness because they have great freedom to respond to each repair job optimally, and the intrinsic challenges of that job are not great

*Passive Jobs.* Passive jobs often result in situations in which job skills gradually atrophy over time because they are not being used. This can happen when people have too many skills and are overqualified for a job -- there is no challenge. It can happen in factory work where highly skilled people have their jobs simplified and automated. In these cases, workers have low control over the job, but also experience low demands. These jobs do not result in much stress, but people engaged in this type of job do not learn much, either.

*High strain jobs.* High stress jobs are those that are very psychologically demanding but also do not allow much control over how to do the job or the pace of work. There are many examples of jobs like these, although a common example is clerical work, for instance, a person who processes medical or insurance claims. A typical work-flow scenario might go like this: open the form, type on computer, staple something to it, file it, grab the next form. The work requires concentration - the worker cannot "let up" or mistakes will be made.

The work process is highly regimented -- with little control over how to do the job, and no control of the flow of work as it comes across the desk. This kind of job can cause stress over time because there is no, or low skills use, no control over how to do job, but it still has the psychological demands of accuracy.

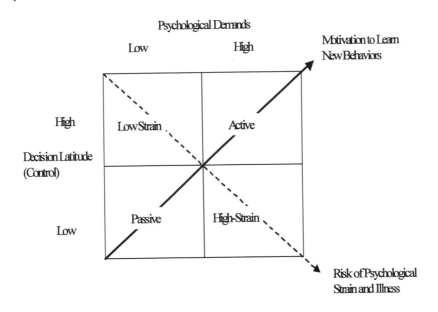

**Figure 3.3** Psychological demand/decision latitude model (Adapted from *Job Demands, Job Decision Latitude, and Mental Strain: Implications for Job Redesign*, by R. A. Karasek, published in Administrative Science Quarterly, Vol. 24, no. 2, by permission of Administrative Science Quarterly.)

This combination of factors is a recipe for a high stress job. This situation is not just limited to "low skill" jobs such as clerical work. Given the right circumstances, such as organizational culture and management style, many professional people face situations of psychologically demanding work paired with fast, unrelenting pace or limited control over the job.

What happens to a person who is continually in a high demand, low control job situation? As we mentioned, these job characteristics cause stress, putting the body in a state of arousal. This arousal is a biological state called the "fight or flight" response. This is the kind of response that results in stories of superhuman strength, such as people lifting cars from trapped accident victims. However, in the case of the office worker, there is no car to lift, nothing to run from (presumably), and no one to fight (hopefully). The body is in a state of arousal but there is no translation into action to make constructive use of those hormones. Instead, the office worker remains seated in his cubicle or meeting room, heart pounding, angry, palms sweating, etc. In this state, the body is producing and releasing various hormones into the bloodstream, including cortisol and adrenaline. Later at home the hormones released earlier are still present in elevated levels in the bloodstream, and blood pressure remains elevated. Over time, this situation is very damaging to mental and physical health, including increased risk for hypertension and coronary heart disease (CHD).

## II. LEVELS OF CONTROL

In this model, control can be exerted from the micro to the macro organizational and physical work environments. Environmental control can be exercised through several mechanisms at each scale of the environment (see Figure 3.2). In this model, we consider three scales of the work environment and organization (see boxes, far left of Figure 3.2) including; Facility/Organization or Business Unit, Collaborative Spaces/Departments/Teams, and Workspaces/Individual. Within each of these levels of organization and space, there are various adaptive mechanisms that support environmental control, and job control (see "Adaptive Mechanisms" column, center of Figure 3.2).

### A. Facility/Organizational Level

At the organizational level, the potential for control can be planned into a building by designing a structurally sound and adaptive facility. The business mission may demand that the building be recycled over time to accommodate new uses. For instance, in Milwaukee, Detroit and other cities that had large manufacturing bases, factories have been converted to retail spaces, condominiums, and modern office spaces. The square footage of the facility may need to be expanded or reduced, and require the internal flexibility to accommodate updated mechanical, computer, telecommunications, and electrical capabilities (Hahn, 1990). The building must be financially viable to continue in operation for an extended period, as opposed to becoming obsolete and forcing the organization to construct a new facility. Building performance can be quantified against the need for environmental control. Hahn (1990) suggests measures such as: replacement factors of major building components, operating cost factors, adaptability costs for changing work uses, and a rating system for site adaptability.

Facility effectiveness can also be measured as a function of how well the building permits the organization to meet broader strategic goals (Salustri, 1990). For instance, large hospital facilities are increasingly becoming customer (that is visitor and patient) focused. Competition has forced these businesses to adopt practices to attract and retain customers. The goal of becoming a customer-focused business from what used to be an industry that was not faced with such restraints has caused hospitals to adopt many new marketing and other practices. Because the introduction of new technologies (and customer services) forces an almost continuous reshuffling of areas within the facility, way finding has become an issue for visitors and patients (O'Neill, 2000). Thus, a strategy emphasizing the ability for the organization to swiftly integrate new services must result in a tactic of maintaining ease of use (especially in terms of "way-finding") of the facilities. The ability to implement a tactic of ease of use is related to the degree of control available by the organization over the facility. Thus, the integration of environmental control into facility design may enhance effectiveness at the organizational level.

If the corporate mission demands continual reorganization around customer or market needs, ease of facility adaptability and reuse will permit changes in space usage to occur incrementally on an as-needed basis, rather than in the context of large, infrequent interior renovations. This type of control could be exercised in the capability to support high "churn" rates. (Churn is usually expressed as a percent, and is calculated by dividing the number of workspace moves by the number of employees in the facility.)

The existence of flexible building design features that provide the opportunity for control constitute a reciprocal mechanism for interplay between the organization and its work spaces.

## B. Group Spaces

The potential for environmental control at the Departmental or Team level is determined by the flexibility of the physical (and organizational) boundaries surrounding the team. Boundaries serve to control the flow of information to and from the group. Control may be exercised by the ability to determine and self manage the reconfiguration of work space layout and boundaries as the current stage of formation of the team, and overall mission, dictate. Organizational boundaries can be reinforced or made more permeable through the design of the physical work space. Boundaries can be used both to integrate the team into the larger organization, or differentiate the team from the organization (Sundstrom and Altman, 1989). Research indicates that the way teams manage interaction across their organizational boundaries influences satisfaction and group effectiveness (Gladstein, 1984).

When the team mission depends on external integration, effectiveness can depend on pace and timing of information exchanges with other units. Team effectiveness can also hinge on the ability to isolate certain activities from outside interference, such as problem solving meetings or sensitive advance R&D areas. The model suggests that both the integrative and differentiating functions can be supported by team control over the layout of the work environment. Control over layout and boundaries may also influence inter-member communication and cohesion. Face-to-face interaction may be influenced by proximity of workstations and meeting places (Sundstrom, 1986).

This model suggests that control over the environment is a mechanism for enhancing the fit between the work group and other organizational elements. The model also proposes that environmental control, as a mechanism for mediating and supporting team member interaction, could reduce stress and problems for team members.

## C. Individual Workspace

The potential for environmental control at the individual level is determined by the adjustability of the physical elements within the work station, such as: task lighting, shelves, storage, work surface height, level of enclosure, VDT screen and keyboard, and HVAC system (Heating, Ventilation and Air Conditioning). Control over these elements may contribute to comfort, environmental satisfaction, privacy, communication, and other perceptions that are related to effectiveness and quality of work life. Many of these features are also related to ergonomic and health considerations (Scalet, 1987). At the individual level, research suggests that environmental control over workstation components has a direct relationship to performance (O'Neill, 1994). A flexible work environment may also enhance job control, by permitting adjustments to the physical environment to be made according to individual task demands or the needs of team workers at a particular stage of a project.

## 1. Research on Control at the Individual Level.

In a field experiment examining the reactions of 400 insurance underwriting and accounting employees, Kroner et al. (1992) found that when workstations were equipped with user-controllable air control units, objective measures of performance increased an average of 2.8 percent. Paciuk (1990) examined the satisfaction and thermal comfort of workers in 511 workstations within 10 buildings. Control was defined as: availability of control (type and degree of control made available by the environment); perceptions of control (knowledge of available control strategies); and exercised control (frequency of engaging in behaviors aimed at regaining thermal comfort). Paciuk found that perceived control was positively related to worker satisfaction.

O'Neill (1992) examined perceptions of the thermal comfort of 200 office workers within a building that had standard HVAC systems on some floors, and underfloor mounted user-adjustable air control units on other floors. The study measured several aspects of control, including: availability of control (access to user-controlled HVAC), exercised control (frequency with which employees manipulated the interface controls of the air system), perceived control, and importance of thermal comfort as a goal. He found that perceived control was a significant predictor of thermal satisfaction, and that exercised control was the most important predictor of employee perceptions of air quality.

Zimmer and Cornell (1990) investigated worker reactions to flexible individual and group spaces. Spaces were designed to give users environmental control through flexibility of lounge furniture, tables, tack boards, white boards, and lighting. They found that 89 percent of participants felt that the flexible elements in the spaces made them more effective at their jobs.

O'Neill (1994) investigated the relationship between environmental control, satisfaction with workstation, and the self-assessed performance of 541 managerial and non-managerial workers in 14 office buildings. Environmental control was measured as: the degree to which the work space supports organization of work materials, ease of adjustment of storage and display features, and ease of rearranging furnishings. Regression and path analyses revealed that control contributed directly to satisfaction and performance, and indirectly to these outcomes through mediating perceptions such as distractions, privacy, and communication. The findings of this study also suggested that workers use partition enclosure features to control privacy and visual access by others into their spaces.

O'Neill (1995a) examined the relationship between job type, workstation design features, and the self-reported psychological stress, musculoskeletal pain, and health of office workers. A total of 62 workers in four job types working in four field sales offices of a U.S. computer manufacturer provided data. Regression analyses tested three indexes of workstation design as independent variables, including: environmental control through adjustability of workstation features, layout of workstation to match job requirements, and appropriateness of storage.

Environmental control was measured through a six-item index including: adjustability and ease of adjustment of seating, ability to rearrange furnishings, adjustability of lighting, ease of organizing work materials, and overall adjustability of workstation. The study reported that for sales professionals, who worked in teams, lack of control over the workstation environment is a significant predictor of psychological stress. For computer technical professionals, lack of environmental control and inappropriate layout were significant predictors of stress and general health. Environmental control predicted 50 percent of the variance in overall health assessments for general office workers.

In a separate study, O'Neill (1995b) reported that for sales professionals, environmental control was a significant predictor of group effectiveness. For computer technical support workers, environmental control was a significant predictor of individual performance and work group effectiveness.

## III. DISCUSSION

This chapter shows that the model of environmental control is supported to some extent at the individual and work group levels by the empirical literature. The concept of environmental control appears to be robust enough to be extended to the organizational scale, as well. While only a few studies have specifically examined the relationships between environmental control and performance or stress, the findings suggest a consistent relationship between opportunities for control provided by design features within office environments for teams and individuals, and gains in work performance and health indications.

Methodological difficulties, such as the need for objective measures of stress and control, and the need for longitudinal designs to detect stress and health effects have contributed to a lack of research in this area (O'Neill and Carayon, 1993). However, the model provides a framework to guide our thinking. At the organizational level, this model could provide the basis for developing new measurements of effectiveness and relating them to organizational strategy and facilities design and planning. We suggest the development and application of environmental control as part of a proactive strategy for reducing stress and enhancing worker effectiveness in organizations. This is part of a perspective shift away from a reactive stance, which emphasizes "damage control" of existing problems (i.e., "wellness" programs to treat stress and health symptoms or training programs to enhance coping skills). The reactive approach may place a disproportionate burden of adapting on the individual. Failure to adapt successfully to the job or to an inappropriate work environment can lead to additional stress (Karasek and Theorell, 1993).

The approach we suggest is proactive, aimed at the elimination of stressors that are the source of these problems. This approach can be integrated as part of a long-term strategy emphasizing the prevention of stress problems through multiple techniques, such as job redesign, employee participation, self-directed learning and, as our model proposes, the integration of environmental control throughout all levels of the organization.

As workers are increasingly required to take an active role in learning and decision making in their jobs, and as the costs of stress and poor health increase to organizations, the facility must respond in different ways to demands for flexibility as individual job definitions change, to facilitate information flow between team members, to address the needs of work groups as boundaries change throughout the course of a project, and to reconfigure technology and service infrastructure as the mission of the organization changes.

Our model considers the active role of the individual, the team, and the organization, and the availability of control as a mechanism to match the environment to ongoing job demands and goals, and thus promote effective work.

## PART II: THE WORKPLACE PERFORMANCE
## MEASUREMENT PROCESS

# CHAPTER 4

## The Workplace Performance Measurement Process

"If you can't measure it, you can't manage it."
R. Kaplan and P. Norton

In addition to providing a place to work, the intention of most office design projects is to address organizational and business objectives. Those objectives may or may not be well-defined, or even explicitly agreed to, by the stakeholders in the new workspace. Once employees have moved into the new workspace, there is rarely a systematic assessment of the success of the design response in terms of the original objectives. Even less frequently is there any formal ongoing assessment of the performance of the workplace and management services within the framework of business objectives. Given the increasing pressure on organizations to leverage all capital investments (including buildings and interior space) in pursuit of business goals, workspace design is playing an ever more critical role in the success of companies.

In this chapter we introduce a Workplace Measurement Process that uses quantitative behavioral and business metrics, data gathering tools, and analysis methods. This model is strongly process-oriented and supports the use of rigorous quality tools and methods. We designed this measurement model to work with the dynamic characteristics of the biological metaphor of organizations and workplaces that we discussed earlier in this book.

Thus the idea of ongoing performance measurement is central to this model. Ongoing measurement creates a "feedback loop," which provides information about the performance of the workplace to those who manage it. This information can subsequently be applied to correct problems with the workplace as they are detected. Problems can be related to the degree to which facility services or the design of the workspace are supporting the right behaviors and work processes needed to help the business meet its objectives. This idea of ongoing "sensing" or measurement of workspace performance is closely related to the concept of "self-managed" behavior that is central to our biological metaphor. In the biological metaphor, the components of an organization (people, technology, processes) sense their own performance and modify their behaviors over time to better meet the goals of the organization. As we discussed, in this metaphor environmental control is the mechanism by which the components of the organization make ongoing, self-managed changes to their forms or behaviors. Thus, our measurement model provides the "sensing" or feedback component that a biological system requires to understand its current state and make corrective changes.

In addition to the Measurement Model, this chapter also includes two Case Studies in which the methods, tools, and processes of this model are employed. Within this chapter, we employ these Case Studies to illustrate the process used by the Measurement Model.

## I. DESCRIPTION OF THE PROCESS

As we have discussed earlier in this volume, the behavior of organizations is "purpose-driven" towards the achievement of specific objectives. Thus, the behavior of the organization and its internal systems is oriented toward achieving the goal the system was designed to meet. The greater the clarity of the organization's goals and objectives, the better the chance that the internal systems (workplace, technology, processes) will be aligned to support those goals. At the most general level, these objectives will likely include business, financial, behavioral, and other metrics. Regardless of the nature of the objectives, by definition they must be measurable. Because the Biological Model incorporates measurable objectives, it is possible to use an analysis approach that links characteristics of the work system (such as; features of the physical office workspace, characteristics of business processes, technology design, etc.) with impacts on business goals at various levels within the organization.

At the start of the process, the Workplace Measurement Process uses a Workplace Balanced Scorecard (WBS) to identify performance metrics related to the objectives of the organization. The WBS helps to form a framework and boundaries for the types of data that will ultimately be collected and analyzed. Once measures are identified, the Measurement Model uses data gathering methods such as surveys and business process analyses, and taps into sources of Human Resource and business data. For analysis, we employ field research methods and Six-Sigma quality tools. For ongoing measurement, we use the Six-Sigma quality process. All of these concepts are discussed in detail within this chapter.

### A. KPI's (Key Performance Indicators)

Once an organization has analyzed its mission, identified all its stakeholders, and defined its goals, it needs a way to measure progress toward those goals. Key Performance Indicators (KPI's) are those measurements. A KPI is an overall area of measurement, such as "customer satisfaction" or "health" or any other higher-level construct. "Measures" are the specific items being assessed that represent components of that metric. For instance, a "Health" metric could be assessed through any number of individual measures, such as; lost work days, doctor visits, discomfort in various parts of the body, or self-reported psychological stress. The methods used to gather data could include a questionnaire, observations, and other data sources such as safety or human resources databases within the organization.

Measures can be employed as leading indicators or trailing indicators. Leading indicators can be thought of as predictors of future trends. Trailing indicators are measures of outcomes or past performance. Continuing our Health KPI example, self-reported psychological stress is a leading indicator of future potential health problems such as hypertension, high blood pressure, or other medical issues. Measures of Lost Work Days and Doctor Visits are trailing indicators of health trends, indicating health outcomes that have already occurred.

Key Performance Indicators are quantifiable measurements, agreed to beforehand, that reflect the critical success factors of an organization. They will differ depending on the organization. A business may have as one of its Key Performance Indicators the percentage of its income that comes from return customers. A school may focus its Key Performance Indicators on graduation rates of its students. A Customer Service Department may have as one of its Key Performance Indicators, in line with overall company KPIs, percentage of customer calls completed without having to transfer the call. A Key Performance Indicator for a social service organization might be number of clients assisted during the year.

Whatever Key Performance Indicators are selected, they must reflect the organization's goals, they must be key to its success, and they must be quantifiable (measurable). Key Performance Indicators usually are long-term considerations. The definition of what they are and how they are measured does not change often. However, the goals for a particular Key Performance Indicator may change as the organizations goals change or as it get closer to achieving a goal.

## 1. Key Performance Indicators Reflect Organizational Goals

An organization that has as one of its goals "to be the most profitable company in our industry" will have Key Performance Indicators that measure profit and related fiscal measures. "Pre-tax Profit" and "Shareholder Equity" will be among them. However, "Percent of Profit Contributed to Community Causes" probably will not be one of its Key Performance Indicators. On the other hand, a school is not concerned with making a profit, so its Key

Performance Indicators will be different. KPIs like "Graduation Rate" and "Success In Finding Employment After Graduation," though different, accurately reflect the school's mission and goals.

## 2. Key Performance Indicators Must Be Quantifiable

If a Key Performance Indicator is going to be of any value, there must be a way to accurately define and measure it. In addition, the manner in which it is measured over time must be consistent. "Generate More Repeat Customers" is useless as a KPI without some way to distinguish between new and repeat customers. "Be The Most Popular Company" won't work as a KPI if there is no way to measure the company's popularity or compare it to others. It is also important to define the Key Performance Indicators and stay with the same definition from year to year. For a KPI of "Increase Sales," you need to address considerations like whether to measure by units sold or by dollar value of sales. Will returns be deducted from sales in the month of the sale or the month of the return?

A company goal to be the employer of choice might include a KPI of "Turnover Rate." After the Key Performance Indicator has been defined as "the number of voluntary resignations and terminations for performance, divided by the total number of employees at the beginning of the period," a way to measure it needs to be set up by collecting the information in the right format from the Human Resource database.

## 3. Selecting Relevant KPI's

Many things are measurable. That does not make them key to the organization's success. In selecting Key Performance Indicators, it is critical to limit them to those factors essential to the organization reaching its goals. It is also important to keep the number of Key Performance Indicators small just to keep everyone's attention focused on achieving the same KPIs. Like the old design adage, "less is more."

That is not to say, for instance, that a company will have only three or four total KPI's in the company. Rather, there will be three or four Key Performance Indicators for the company and all the units within it will have three, four, or five KPIs that support the overall company goals and can be "rolled up" into them. In addition, KPI's developed for use in measuring workplace performance by the Real Estate or Facilities team must be selected on the basis of some potential, logical link between what is measured and the potential impact that physical workplace design and facilities management practices and processes might actually have on the KPI.

For instance, if a company-wide Key Performance Indicator is "Increased Employee Retention," the Real Estate or Facilities team measures in support of that KPI could logically include that KPI because it is possible that employee satisfaction with various qualities of the workplace, and services provided, could be related to their intention to stay with the company.

## B. Workplace Balanced ScoreCard

In our approach, we organize the KPI's in the format of a "Workplace Balanced Scorecard" (WBS). Each heading within the WBS represents a "Metric" or KPI (see gray headings, Figure 4.1). The KPI categories selected and the emphasis or importance placed on each will vary according to the company's industry, market, and strategic objectives. The Workplace Balanced ScoreCard approach is based on the balanced scorecard approach and methodologies described by Kaplan and Norton (1996). There are some critical differences between Kaplan and Norton's Balanced Scorecard (BSC) approach and the WBS. The BSC is a comprehensive organizational and management tool designed to drive overall organizational excellence through optimized management practices and ongoing measurement and feedback of performance data towards business goals. It is a philosophy of management and a reflection of organizational culture.

Our WBS is a tool and methodology intended to result in KPI's and measures pertinent to the potential contribution or influence that the facility and office workspace might have on organizational goals and business KPI's already in place within the organization.

The WBS itself is intended to be a tool for the strategic management of the facility, office workspace, in the context of broader organizational objectives. Thus the KPI's generated for the WBS should be related to and will likely overlap with higher-level organizational goals. They should certainly at a minimum be congruent with organizational KPI's and objectives. For instance, suppose an organizational KPI is for a company to be "the innovation leader" within its industry. One example of a supporting WBS KPI might be collaboration. Many managers believe that increased collaboration has many benefits such as speed of decision making and higher quality work outputs of teams and business units. Thus, level of collaboration could be logically related to increased innovation. The design of the workspace could logically be optimized to enhance collaborative behaviors.

Thus, WBS KPIs should be more directly related to contributions that the workplace makes in supporting behaviors and business processes. Like the original Balanced ScoreCard, the WBS is intended to form the basis of an ongoing measurement program that provides feedback to managers. However, the WBS is also designed to assist managers in optimizing the performance of the workspace against the KPIs. When coupled with an ongoing measurement program, managers can use the WBS and the Workplace Measurement Process described in this chapter to maximize the strategic alignment of the office space with objectives relevant to the business success of the organization.

The KPIs shown in Figure 4.1 are examples that Real Estate and Facilities/Services groups have used. These examples include KPIs such as "Market Leadership/Customer Needs," KPIs pertaining to the quality and effectiveness of the workspace and desired behaviors, "Work Environment/Behavior," wellness issues within the "Health" KPI, and on internal financial impacts "Financial/Business Process Efficiency." The WBS and its KPIs can be used by an organization in a variety of ways: to influence the direction of overall design guidelines for a new building, to inform workspace standards; to provide the metrics for a workspace evaluation; or (ideally) to form the basis of an ongoing workplace quality measurement program. In any case, what is measured should be relevant to the business KPIs for which the senior management team is responsible. When creating a WBS, a person who can represent the management team should be a part of the review process in order to solicit input and help make decisions about which KPIs to include in the WBS.

The individual measures "underneath" each KPI are shown under each of the KPI headings (see gray heading boxes, Figure 4.1). In practice some subset of each of these measures would be identified and used in a measurement program. Ideally, the smallest number of measures will be used that will have the greatest potential impact on the organization. This is important for a number of reasons, the least of which is that every additional measure adds cost and time to a measurement project. This is especially true if the measures are implemented within the context of a permanent workplace quality assessment program, in which measures would be taken at regular intervals. In addition, the greater the number of measures, the more complex the reporting and the findings are to understand and interpret - particularly for those outside the immediate measurement activity, who will need to understand and take action on the results.

Having fewer measures also decreases the chances of errors that might result when manipulating and analyzing complex data sets. Fewer measures will lead to greater simplicity in the analysis of data and increase the chances that management will appreciate and understand the story you are telling about the results.

## C. The Workplace Measurement Process

The Workplace Measurement Process is a process and methodology that ultimately allows us to connect the impacts of workplace design features (from small to large scale) to changes in measures of behavior, performance, health, and business process, using the KPIs in the Workplace Balanced Scorecard (see Figure 4.1).

The Workplace Measurement Process has five phases of activity (see headings on Figure 4.2). This approach permits the creation and measurement of KPIs at three different scales, or levels of analysis: enterprise-wide, business unit (or department), and group/individual (see triangle at left of Figure 4.2).

| Financial/Business Process Efficiency | Market Leadership/Customer Needs |
|---|---|
| - **Employee Attraction** (# open job requisitions, time open) <br> - **Business Process** (time and cost) <br> - **Merit Review Scores** <br> - **Overtime Costs** <br> - **Call Center Agent Performance** (call transfer rate, customer ratings) | - **Time to market** <br> - **Innovation** (new products per quarter, patents filed, number of R&D projects) <br> - **Customer Satisfaction** |
| **Work Environment/Behavior** | **Health** |
| - **Fit of workspace design features to work needs** <br> - **Support for Collaboration** <br> - **Sense of Community** <br> - **Support for Group work** | - **Lost days** <br> - **Medical Claims** <br> - **Discomfort** <br> - **Psychological Stress** |

**Figure 4.1** Example of a Workplace Balanced Scorecard (WBS) (Author)

In the first phase of the process, KPIs are defined and baseline measures of existing conditions are made, (see "Define KPI's" column, Figure 4.2). Once the changes to the workplace are made (see "Workplace Redesign" column, Figure 4.2), analyses of the impact of the design intervention on those metrics (see "Analyze Impacts" column, Figure 4.2) can be made.

The next step in the process is the "Improve" phase; in which the data is interpreted and acted upon using processes such as "Root Cause Analysis," the creation of a Response Plan to address issues; and "triage," a way of prioritizing problems and actions (see Figure 4.2). These Improve activities occur regardless of the level of the organization at which the workplace has been changed; thus Figure 4.2 shows these activities cutting across all three levels of the organization. Each of these Improve activities is discussed in greater detail later in this Chapter.

In the "Control" phase of our Workplace Measurement Process, we seek to "hold the gains" made by the improvements to the workspace (see far right column, Figure 4.2). Activities within the Control phase may include implementation of a permanent workspace performance tracking system, using Six-Sigma tools such as Control Charts to interpret trends, sharing best practices with colleagues in different parts of the company or through a network of peers in other organizations, or sharing information through a formal internal workplace communication program. In general, the intent of Control activities is to assist the team in sustaining improvements and communicate what has been learned to various parts of the organization.

The bottom of Figure 4.2 shows an arrow (moving from right to left) representing a feedback loop from the Control phase back to the Business Objectives at the far left of the diagram. This feedback loop illustrates the strategic value of using the analysis to determine the success to which the workspace and continuous improvement activities meet stated performance goals. The feedback loop, which emanates from the Control phase, also represents the notion of continuous improvement - where the information gathered is also applied to make changes and improve the workplace and supporting services. The use of quality programs and continuous improvement is discussed in more detail in the following sections of this Chapter.

For instance, at the company-wide (enterprise) level, the example Workplace Measurement Process shown in Figure 4.2 shows "Employee Retention" defined as a KPI. In the next column to the right, the response to that issue is shown - which in this case is to use the space to effectively communicate corporate image and sense of belonging. Keep in mind that there may be other, non-workplace design responses also in place or being implemented, such as benefits/compensation plans, or other management or HR programs. In the Analyze phase (next column to the right, Figure 4.2) the effects of the workplace redesign are assessed by measuring voluntary separations (and probably other measures as well). The measures shown under the "Analyze Impacts" column are the measures developed under each KPI. It is likely that there would be more than one measure under each KPI. In the Improve Phase, this information would be used to formulate a response in terms of changes to the work environment or supporting services to address any needs related to retention. In the Control phase, the information regarding retention could be tracked over time, and the responses shared with colleagues and internal customers through the communication plan.

The Workplace Measurement Process helps to create a line of sight between business objectives and workplace design decisions. When design concepts are being developed and evaluated, it can help keep the team on track in terms of having clearly articulated objectives and metrics, so that there is a rationale and defense of design decisions. It brings a clarity to project objectives that the team can agree to and makes communicating those objectives to internal customers easier.

Once the project is complete, the data collected can be organized and presented to management and other stakeholders in the workspace in the form of "proof statements" of project success that use the agreed-upon metrics.

Finally (and from our perspective, most importantly) the Workplace Measurement Process can form the groundwork for implementing a system of continuous improvement of the workplace. This is a critical aspect of the measurement model, since the Biological System is a dynamic entity that changes over the course of time, which results in misalignment of the office workspace with the objectives of the organization. Measurement of workspace performance at regular intervals - along with structured responses to problems - is an ideal process to support high-performing office workplaces.

## D. The Workplace Measurement Process and Six Sigma

Because Six Sigma is rooted in improving manufacturing processes, not all aspects of this approach neatly fit the needs of facility design and management. However, the key principles of Six Sigma - that of identifying key performance metrics, collecting and analyzing data on an ongoing basis, and the use of that data as a management tool for continuous improvement of work environments - remains central to our Workplace Measurement Process.

### 1. What is Six Sigma?

Six Sigma is a statistical measure of the quality of products or services based on studying the variation in those products or services as they are created or delivered. For instance, in manufacturing, as widgets come off an assembly line, the variation in a key dimension of the widget can be recorded and tracked. Six-Sigma tools track the variation around the mean size of that part of the widget. The assumption is that variation in the parts reflects underlying process variation.

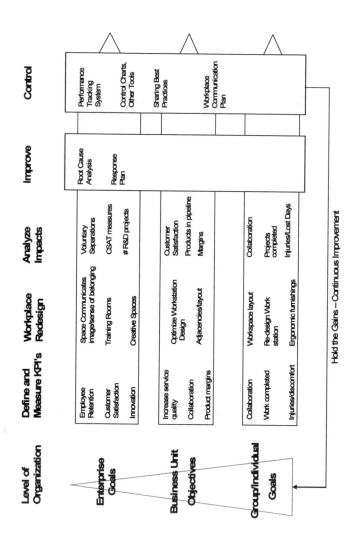

**Figure 4.2** Workplace Measurement Process (Author)

If process variation can be reduced, then the variation in widget size can be reduced too. In workspaces, variation in customer evaluations of aspects of that work environment (such as comfort, support for communication, privacy, etc.) can also be tracked using Six-Sigma tools.

Variation, when measured statistically, is the standard deviation around the mean, represented by the Greek symbol, Sigma (Eckes, 2001). To achieve Six Sigma quality, a process must not produce more than 3.4 defects per million opportunities. The statistical representation of Six Sigma describes quantitatively how well a process is performing, whether that process is in support of creating widgets on an assembly line, or evaluations of a workspace. A Six Sigma defect is defined as anything outside of customer specifications. While the reduction of process variation is a focus of Six-Sigma activities in manufacturing, another central component of this approach is *process improvement*. In manufacturing, process improvement should result in reduced variation in the required dimensions of widgets. In management and design of office workspaces, process improvement can be measured through improvements in internal customer satisfaction with the space, or evaluations of specific features of the space.

Six Sigma improvement projects (George, 2002) reduce variation and improve products and services using any number of existing Six Sigma methodologies. Often companies modify existing Six-Sigma processes to fit their own requirements. One commonly used Six Sigma process is called DMAIC (Define, Measure, Analyze, Improve, Control), which improves existing processes and products (George, 2003). We have adapted the general DMAIC process to become a core part of our Workplace Measurement Process (see Figure 4.2). We applied the general principles of this approach within our model because it provides a structured but flexible set of steps or phases to measure and improve workspaces and support services.

To make this discussion more concrete, we describe how we applied our Workplace Measurement Model with an organization that wanted to measure and improve their office facilities.

## E. Case Study 1: Workplace Measurement Process for Evaluating and Improving Workspaces

This company is a widely diversified leader in the consumer products industry. A household name, its business strategy requires innovation by promoting behaviors such as collaboration, communication, group work, and mobility. To support that strategy, they intended to create flexible spaces that support the desired behaviors, encourage a sense of belonging, and communicate corporate image to employees and customers. They felt that their current workplace standards no longer fit their strategy.

The organization implemented a Pilot Project to test ideas about changing the way workspace is allocated, planned, designed, and managed. The pilot program was viewed as a "working lab" to test the elements of the design strategy and to apply the learning for a new Headquarters building. They also wanted to ensure that the space would support the new ways of working for employees within all job types, from executives through clerical staff. To provide credible information to support the pilot and improve the workspace, we applied our Workplace Measurement Model.

## 1. Define Phase

We formed a Steering Committee that included the VP of Real Estate and Director of Facility Services to provide guidance and support to the Project Team over the course of this project. The Project Team consisted of members of our Consulting team, the (external) Design Team, and other members of the Real Estate and Facility/Services groups. The first task of the Project Team was create a Workplace Balanced ScoreCard (WBS) through a facilitated work session in which KPIs (Key Performance Indicators) were identified, based on the Vision Statement for the organization.

a.          Identify Business Drivers -- The team broke into small groups and created lists of driving and restraining forces to the business that lie in the present and immediate future. These form the context for identifying and creating the KPI's within the Workplace Balanced Scorecard. A sample of this list is illustrated in Figure 4.3.

b.          We discussed the concept of the Workplace Balanced Scorecard (WBS) and how it can create general areas of Measurement (KPI's). The KPI's are the general areas of "what" will be measured.

c.          To get the discussion started in identifying KPI's, the team brainstormed a list of general areas, then used a simple voting technique and further definition to reduce the list to four KPIs: Financial, Market/Customer, Work Environment/Behavior, and Health (see Figure 4.1).

These form the headings, or Metrics used in each of the four quadrants within the Workplace Balanced Scorecard. The team also created a definition for each KPI. For example, for the Market/Customer KPI, the definition was "To meet our goals around Markets and Customers, what behaviors and business processes must the office workplace support?"

The KPI's were reviewed with the Steering Committee. During that time final project plan and time lines were created using a Gantt chart and agreed to by the Steering Committee.

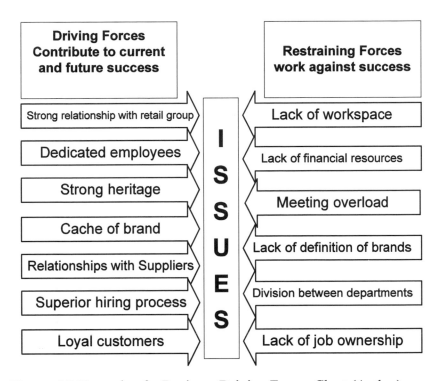

**Figure 4.3** Example of a Business Driving Forces Chart (Author)

d.          The next step was to create the list of potential measures for each KPI within each Workplace Balanced Scorecard quadrant. The team broke into small groups and brainstormed a list of as many measures as possible underlying each KPI, ultimately identifying 31 measures for all the KPI's.

e.          Once the list of measures was identified and a definition for each was agreed upon by the group, participants were asked to evaluate each metric on two dimensions: 1) the relative importance of having each metric to the future success of the organization; and 2) estimated difficulty in terms of time and cost to measure and report each metric. The voting is a confidential process using a personal voting terminal connected to a proprietary software package that instantly analyzes the results (see Figure 4.4).

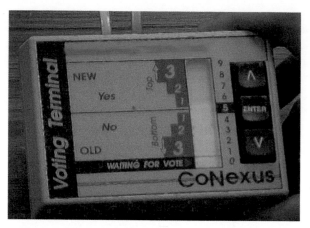

**Figure 4.4** Voting Terminal (Author)

f.          The results of the group vote were instantly displayed for the group to see in the form of a four-quadrant matrix "opportunity profile" that plotted the future importance of each measure against its estimated difficulty of implementation (see Figure 4.5). Figure 4.5 shows the keywords representing each measure and their location within the four areas within the "Profile of Results."

The group reviewed the results and selected a subset of metrics that represented a balance between items that would strongly contribute to future success, with the time, cost, and difficulty involved in creating a system or process to collect measures associated with that metric.

g.          Once all the measures were ranked, the two highest rated measures (in terms of importance and difficulty) for each KPI were placed into the Workplace Balanced Scorecard quadrants.

h.          The final step in creating the Workplace Balanced Scorecard was to identify the sources of data for each of the measures. If a source of data for a particular measure was not available, the team identified a method for collecting the required data (survey, observations, or other methods).

Profile of Results

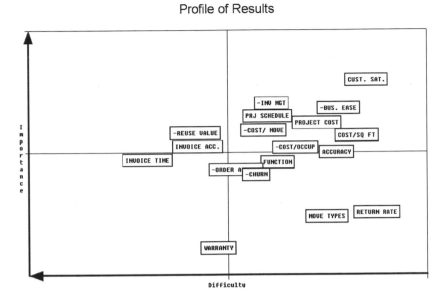

**Figure 4.5** Profile of Results from voting (Author)

The detail on the individual measures for the Work Environment/Behavior KPI focused on three areas of employee behavior and perceptions: Employees' Sense of Belonging to the Company, Collaboration, and Quality of internal group processes (shown in Table 4.1).

| Table 4.1 Individual Measures for Work Environment/Behavior KPI |
| --- |
| **Sense of Belonging to the Company** |
| -This office space conveys the appropriate image of the company to employees and others. |
| -The workspace helps team members feel like they are really part of the team through design features or visual cues. |
| -The design of the space near my workstation contributes to my sense of belonging to the organization. |
| **Collaboration** |
| -The overall workspace lets me quickly shift from individual work to collaboration with others. |
| -The design of the various spaces in this office provides adequate support for collaboration. |
| **Quality of Internal Group Processes** |
| -The workspace supports team member participation in the ongoing work. |
| -It is easy to physically access co-workers when needing to discuss a work issue. |

## 2. Measure Phase

In this phase, we finalized the specific measures to assess the KPI's (see Table 4.1). Throughout the study, we collected data at regular intervals from both experimental and control groups (Keppel and Wickens, 2004; Coleman, 2002). The Experimental group received the new workspaces; the Control group did not receive any changes to their workspaces. The study involved 180 employees within two Pilot Phases. In Phase I employees were moved into a new space designed to enhance performance on the Key Performance Indicators. In Phase II the design of the Pilot spaces were fine-tuned based on feedback after the first phase. The data collection tool was a questionnaire, developed with a series of questions that tapped into each of the KPI's, as well as other issues related to workplace design performance.

This was deployed within the idTools® survey analysis system used by the team to collect the data and provide the initial analyses (see Figure 4.6).

## 3. Analysis Phase

Once we collected the data on key performance indicators, we employed a Six-Sigma method, Statistical Process Control (SPC) to analyze the process variation measures on each of the three measures within the Work Environment/Behavior KPI. The primary tool of SPC is the Control Chart, which permits identification of the sources of process variation (discussed in detail in the following section). The Control Charts used for this project were generated through the MiniTab® Statistics software package -- an inexpensive and commonly available application.

## 4. Statistical Process Control

Using SPC tools, the variability in a process over time (in this case the quality of a work environment) is monitored by collecting data at regular intervals. Data is displayed and analyzed through Control Charts, which can be used to track results and reveal whether adjustments need to be made to the process (or to the work environment) to reduce the variability and bring the process back "into control" if needed. Control Charts can determine if process improvement efforts (such as with workplace design) are having the desired effects.

Two types of variability can occur in a process (or work environment) that is being tracked - common cause variation, or special cause variation (Eckes, 2001).

| | | Standard Report | |
| --- | --- | --- | --- |

Back

**Filter Selection**

| GU | Country | Location/Building |
| --- | --- | --- |
| ALL<br>ASG<br>Canada<br>Gallia | ALL<br>Austria<br>Argentina<br>Australia | ALL<br>Akasaka - Dai-2 Bldg.<br>Akasaka - DS Bldg.<br>Akasaka - Intercity Bldg. |

| Workforce | Level | Average time spent in the Building |
| --- | --- | --- |
| ALL<br>Consulting / Bus. Consulting (BC)<br>Consulting / Bus. & Syst. Int. Consulting (BSIC)<br>Consulting / Technology Consulting (TC) | ALL<br>Analyst / Sr. Analyst<br>Analyst Programmer / Sr. Software Engineer<br>Assistant / Sr. Assistant | ALL<br>Less than once a month<br>1- 4 times per month<br>5 - 10 times per month |

**Report Selection**

| Background Information | Workplace & Work Environment | Office Services |
| --- | --- | --- |
| Travel Services | Administrative Capability | Executive Support |
| Best Practices and Suggestions | | |

| | Workplace & Work Environment | | | | | | | | |
| --- | --- | --- | --- | --- | --- | --- | --- | --- | --- |
| Question | # | Avg | Std Dev | NA | 5 | 4 | 3 | 2 | 1 |
| Overall, this location provides an engaging workplace [appropriate image of Accenture, welcoming environment, collaborative workplace]. | 1404 | 4.03 | 0.85 | 87 | 396 | 777 | 136 | 72 | 23 |
| The types of workspace [open, enclosed, touch down, etc.] I need to be productive | 1395 | 3.87 | 0.89 | 96 | 292 | 788 | 175 | 120 | 20 |
| Appropriate conference and meeting spaces available for my needs | 1324 | 3.81 | 0.93 | 167 | 267 | 721 | 187 | 122 | 27 |
| A simple and effective process for reserving workspace/meeting rooms | 1300 | 3.93 | 0.89 | 191 | 316 | 706 | 162 | 99 | 17 |
| Efficient printers, copiers and fax machines | 1384 | 3.88 | 1.02 | 107 | 388 | 660 | 149 | 155 | 32 |
| Reliable tools for holding virtual meetings [conference calls, video conferencing] | 1116 | 3.86 | 0.82 | 375 | 205 | 637 | 202 | 56 | 16 |
| An appropriate level of safety & security | 1340 | 4.05 | 0.82 | 151 | 368 | 769 | 124 | 61 | 18 |
| A clean environment [building and workspaces] | 1418 | 4.26 | 0.85 | 73 | 626 | 634 | 86 | 47 | 25 |
| Total Mean | | 3.97 | | | | | | | |

Show Comments

Figure 4.6 idTools Report screen (Author)

Common cause variation is simply the normal variation about the statistical mean that occurs when data is collected at many points over time. Special cause variation comes from outside events or fundamental problems with the process (or in this case, with which the design of the office workspace) in terms of meeting employee needs.

The project team's goal was twofold: to reduce special cause variation and to improve scores on the KPI's over time. In the case of work environment evaluation, special cause variation would be due to lack of fit between employee needs on the KPI's and the design or management of the space in support of those needs. Figure 4.7 shows the results of the project on one KPI, "Sense of Belonging to the Company." Due to space constraints we will focus on this example to fully explain the aspects of this control chart and how the data applies to this Case Study.

The XR Bar Control Chart is a powerful tool that shows process variability and trends, and permits interpretation of data so that changes can be made to the workplace. The X Bar Chart (entitled "Sample Mean," upper chart, Figure 4.7) shows the Means (averages) of the samples at each observation point over time. The R Chart (entitled "Sample Range," lower chart, Figure 4.7) shows the range or variability in each set of scores at each observation point. In both charts, the centerline shows the Grand Mean (Mean of the Means), about which the scores are plotted. The system calculates boundary lines 3 sigma limits above and 3 below the chart centerline (Mean of the

Means). These charts can then be used to determine if a measure being tracked is within Six-Sigma limits, and is thus "in control."

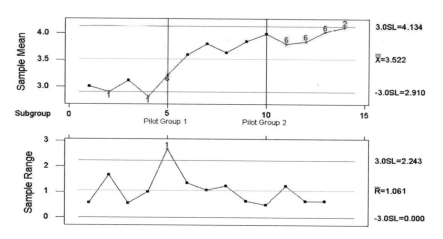

### Sense of Belonging to Company

**Figure 4.7** XR Bar Control Chart (Author)

In a situation in which we are trying to reduce process variation over time (and keep the process under control) and increase scores on a measure, the desired result is to see an upward trend in scores within the "X Chart" (Sample Means Chart) and a downward trend, or scores that cluster tightly about the center "average" line, for the R (Range) chart.

In this study we collected data every 3 weeks from a sample of 20 employees in both Experimental (those who received the new work environment) and Control (those who did not experience any changes) groups.* Figure 4.7 shows that we collected data at 15 points in time over the course of this project. The vertical lines through the X Chart (upper chart, Figure 4.7, "Pilot Group 1" and "Pilot Group 2") show the points at which we implemented design changes to the workspaces.

***Process Control Analyses.*** As mentioned earlier, Control Charts permit the analysis of data collected over time to help assess whether a process is "in control" or "out of control." By definition, when a process is in control, it is within Six Sigma limits. On a given Control Chart, these limits are calculated from the data and are indicated by the "+3.0 SL" (sigma limit) and lower "-3.0 SL" lines. When trying to understand if a process is in

---

\*      The data was analyzed using the Control group so that the true effects of the workplace on this measure could be assessed.

control, both the X Chart (means) and R Charts (range) are evaluated. The Control Chart itself also provides interpretations of the data to aid in assessment of process control. This interpretation is indicated by the numbers associated with several of the data points shown in the example in Figure 4.7.

Each of the tests for "special causes" detects a specific pattern in the data plotted on the chart. The occurrence of a pattern suggests a special cause for the variation, one that should be investigated. When a point fails a test, it is marked with the test number on the chart. For example, the Sample Means being tracked (top of Figure 4.7) show two number "1's" prior to the move of Pilot Group 1 into the new space. This indicates that a special cause is driving the means to be more than 3 sigma's from the center line. In this case we interpret the special cause to reflect a fundamental problem with the design of space that fails to create a sense of belonging to the company. The next two "6's" on the control chart show that another type of special cause variation has been detected, in this case, 4 out of 5 of the points following the first "6" are at least 1 sigma over the center-line. In this case, we are seeking process improvement (increases over the mean). While there is special cause variation underlying this set of scores, this variation above the mean is desirable, and the "special cause" is probably the improved workspace that employees experienced, which is driving the upward trend in scores. The final special cause test, shown by the number "2" on the final data point of Figure 4.7, indicates that 9 or more of the previous data points lie above the center-line. In this case, because we are seeking process improvement and desire scores that are above the mean, this confirms that a "special cause" is at work in the consistently high scores. We assume because of the nature of our project that the special cause variation from the mean is due to the continued improvements to the workspace driving the enhanced scores on "sense of belonging" to the company. These analyses highlight the usefulness of Control Charts as a tool to identify special cause variation.

In traditional Six-Sigma quality/process improvement projects (in a manufacturing environment, for instance), the desire is to hold sample mean scores tightly around the Mean of the Means (centerline) and use these tests to identify when an underlying "special cause" is driving variation in scores over time. In such a case, the identification of a special cause indicates that some undesirable change to the process (a problem with a machine tool, an ongoing operator error, etc.) is causing the variation. The variation is not due to normal or expected small variations in the process due to sampling errors or other normal variation.

In the services (as opposed to manufacturing) context in which we are applying these quality tools, we are looking to improve what are essentially customer satisfaction scores with several aspects of the workspace. Thus these test scores indicate desired "special cause" variation, because they occur just after each of the workplace design interventions. (For a more detailed understanding of the various tests for special cause variation, we refer the reader to Wheeler and Chambers, 1992; Wheeler, 1995).

Although these initial results were positive, the Project team decided to examine the data, combined with observations (walkthroughs) and interviews with people using the new space, to determine whether additional modifications to the space could further improve the KPI score. A second Root Cause Analysis was conducted with this new information that yielded further insights into the design.

The team found that while the space contributed to an enhanced feeling of belonging overall, employees within some job categories were actually feeling somewhat alienated by the new space, which was holding down the overall scores and increasing the variability of the measures. These administrative and support employees were located somewhat at a distance from the new team spaces, and had high levels of enclosure through panel height. While the original model for the space had equally distributed team members throughout the floor plate, the new model centralized their workspaces and team spaces by job function. Thus, in the new space the administrative spaces with high enclosure were clustered together. Because of their location and high level of enclosure these employees felt somewhat disconnected from their teams and perhaps from the company.

The Project team acted quickly to relocate administrative employees' workspaces closer to the teams they supported and lower the amount of enclosure in their workspaces.

Once these changes were made, the data collection continued at regular intervals from both the experimental and control groups (indicated by the vertical line above "Pilot Group 2," data collection points 10 through 15, Figure 4.7). The XR Bar Charts reveal a significant improvement in employees' sense of identification with the company after this change (see Figure 4.7). The X Bar (Sample Means) Chart shows that the "sense of belonging" measure increased above the average (double bar X -- middle line) and stayed above that average line consistently for the 5 data-collection points. The R Bar Chart shows a reduction in the ranges of scores within each data sample during that time period, indicating improved consistency in employees' perceptions of the space.

## 5. Control Phase

Typically, once a facility redesign has been implemented and employees re-located into the new or renovated space, it is usually time to move on to the next project, or "put out the next fire." Thus, the opportunity to learn from the completed project, and to make ongoing improvements, is lost. In the Workplace Measurement Process, we attempt to "hold the gains" made through the project, by use of a Control Phase. The key components of this phase are the implementation of a monitoring plan, creation of a response plan, transfer of ownership (project closure), and sharing the "lessons learned" with the organization.

As part of the monitoring plan, we implemented the idTools application (survey and reports from the database) to help create a permanent feedback loop to gather information on the performance of the work environment. The information collected is a source of data for the Control Charts, which are used for assessing the ongoing quality of the work environment. The team is using these tools to collect data to track the variance in performance on each of the KPI's over time.

The Response plan is a checklist or process by which the facility management team can react to the analysis provided by the Control Charts when a problem is revealed in one of the performance metrics. In this case, the Response plan is a simple set of steps (the Root Cause Analysis) that allows the team to identify and rank any problems, and brainstorm design responses as required. In the case of this project, one of the team members on the original project is also on the Facilities staff. This member was trained on the use of idTools and Minitab and will "own" the ongoing data collection/ analysis activities within the Control Phase, as well as communication responsibilities.

Finally, the data being gathered was shared with the organization through meetings involving leads of the Facility Management staff and tools including a Power Point presentation and a case study. These communication materials focused on best practices and lessons learned and are being considered by another team that is creating standards for use in other parts of the organization.

## 6. Summary

In this case study, we discussed an example of how the Measurement Model, with its associated tools and processes, was deployed to measure and improve one particular KPI. We described the application of the Workplace Measurement Process with a consumer products company in which employees were being shifted from traditional cubicle workstations to a more open design concept. Among other goals, management wanted to use the design of space as a tool to communicate corporate image and foster a sense of belonging to the organization among its employees. This was accomplished through a Pilot Project in which design concepts were developed and refined through feedback from Six-Sigma measures, prior to a wider rollout of this design strategy. Once the pilots were conducted and the assessment complete, the processes and tools were left in place to create an ongoing quality program to track workplace performance against these KPI's.

The team wanted an in-house capability that would put them in control of the evaluation process -- and the opportunity for continuous improvement in the work environment -- without the constant requirement for outside resources to support the work. The team selected the idTools system, an enterprise survey data analysis application for facility management. The team also began the process of tracking this data-using Minitab, which contains Six-Sigma quality tools like control charts.

## 7. Lessons Learned

This project represents a holistic example of the application of the Measurement Process using Six-Sigma tools like Control Charts. This approach does bear valuable results without the added complexity of a control group and the additional required statistical analyses. The simple survey data collection methods (idTools) and the Minitab Control Charts were adequate to perform the analyses and interpretation.

We found that to increase the buy-in and ultimate success of such a program, communication about the program scope and benefits to all sponsors and participants is critical. We suggest spending time up front to create a "story" about the project, activities and potential (strategic and financial) benefits, and links to other Quality initiatives. This story should be communicated in the months prior to beginning the project. Set up a Steering Committee to sponsor and guide the project. If you find it difficult to engage management participation in a Steering Committee, then there might be something lacking in the story you are trying to tell or in articulating the benefits of the project.

Apply the "so what" test to each KPI. When formulating the KPI's (metrics areas) for the project, think carefully about how each metric would ultimately be applied to justify a change in the design of the space or facility management process used to support the space. If you cannot easily see the link between a KPI and specific actions, reconsider the use of the KPI.

The entire process we have articulated in this case study represents a proactive approach to managing the quality of work environments. If your team or corporate culture resists such an approach, consider the need for a more formal workplace change communication program to increase the chances for success.

Finally, treat the process of collecting feedback on the work environment, learning, and change as a process of continuous improvement, not as a finger pointing exercise for assigning blame for mistakes. The team should be aware that the workspace presents an opportunity for continual realignment of design features with the goals and aspirations of the organization.

## F. Case Study 2: Workplace Measurement Process for Evaluating and Improving Facility Management Services

A global management consulting company conducted this project. Their mobile workforce is supported by a high degree of "self-enablement" (tools and technology to support the use of the space, reservations, etc.) and a flexible work environment. Offices provide a variety of workspace allocated on a hoteling model with a population-to-seat ratio of approximately two to one. The offices focus on operational efficiency and maximizing human performance through delivery of effective services to users of the facilities. A cross-functional F&S (Facilities & Services) "Connecting to the Customer" team was created to define customer service objectives and metrics and to create a program to measure F&S performance against these objectives.

This case study describes this consulting firm's efforts to create a North American-wide capability to gather, analyze, and interpret customer satisfaction (CSAT) data within a Quality framework from over 23,000 employees per year. The F&S team collects CSAT data on a quarterly basis, which can be sorted by Location, Job Classification, Department/Business Unit, and other variables.

Local site managers and others use this information to assess their performance and compare against North American averages in each service category, using a variety of simple tools to make collecting and tracking data from the idTools system an easy task. High-scoring locations are asked to share best practices with other locations. In this case study, we used samples of the data to create Six-Sigma "X Bar/ R" control charts, which provide information to the U.S. team on the variability of the processes underlying the services provided.

### 1. Define KPI's Phase

In 1997 the consulting firm began a transition from a traditional workspace environment of "owned" space to a "hoteling" model. Visioning exercises predicted a major increase in the capability for remote connectivity with the potential for employees to help themselves in terms of services and capabilities (called "self-enablement") and to support flexible work in terms of time and location. These foundational shifts in the workplace created opportunity for greater efficiencies in space.

The Facilities and Services (F&S) team achieved significant results, reducing office space in the U.S. 35 percent from 3.1 million sq. ft. to 2.0 million in the period 2001 to 2004. However, the primary data used to evaluate real estate and facility decisions has traditionally been quantity and cost of space. To insure protection and enhancement of their most valuable asset (people) the F&S team needed to understand the impact of real estate and space strategies on human performance. Credible metrics were required to balance against well-established real estate and financial measures.

The workplace metrics that the team eventually developed were based in part on the results of a research project conducted at the Dallas location a few years earlier. This study validated that a transition to a new workplace model with the right design, hoteling, and flexible work processes produced improvements in collaboration, job control, ergonomics, business process time, and other outcomes.

Despite this research, no ongoing data was being collected to produce scalable measures for all their locations. It was essential to know which solutions were most successful for their employees and to measure the key elements of facility and office operations that impact workplace performance. In an environment of continuous change, one-time measurements were insufficient to assure maintenance of minimum standards and continuous improvement over time.

Thus, a "Connecting to the Customer" team was formed to identify key business metrics and implement an ongoing quality improvement system for the workplace. This team decided that customers would define the workplace elements most important for their performance and productivity. Those would be the factors included in the survey and subsequently measured and reported. The customer survey was to be one component of the entire program, the goal of which is to insure proper alignment of facility services and support with customer and organization priorities -- the "Customer" component of a "Balanced Scorecard" (Kaplan and Norton, 1992).

***Principles for Defining Metrics.*** As part of this process, the survey content is periodically screened to insure the elements included are those defined by customers as important. Capability and operational leads who propose survey content may have that content rejected, if it is not among those elements identified by the customers as important. The F&S team also strives to keep the questionnaire relatively short to minimize time demands on employees. The measure of success in terms of survey length is survey response rate, with a target of 20 percent. Each business quarter an additional and different 25% of the population is surveyed. This insures each customer receives the survey only once a year, reducing the likelihood of rejection. In practice, response rates average over 30 percent. When location managers need more detailed information from customers, they employ focus groups and interviews -- rather than adding questions to the survey. The metrics identified by the team included:

- •Location Creates Community
  - oWorkplace image, design, activities, and location convenience
  - oSafety and security
- •Quality of Office Services
  - oTimeliness
  - oProfessionalism
  - oMeeting expectations
- •Tools and Self-enablement
  - oFunctionality of technology and office equipment
  - oTravel services
- •Workspace functionality and availability
  - oAbility to reserve space
  - oDesign meets needs - Right types and proportions of space
  - oErgonomics, privacy
  - oConference and meeting spaces fit needs
- •Administrative Capability
- •Quality of Administrative Support
  - oCustomer gets required level of support
  - oStaff has appropriate skills knowledge and flexibility
  - oCustomer is involved with decisions involving admin support

2. Analyze Impacts Phase

As part of this Phase, the team communicated this survey program to the field, gaining understanding and buy-in from location managers and others. Measuring service performance through customer input, while intellectually accepted, is intimidating and can create apprehension among stakeholders. Significant effort and time was required to reach common understanding of language and process and to socialize the concept across different locations and through staff with varied backgrounds. Constant leadership assurance was necessary to support this initiative. Some of the issues the team grappled with included:

### *From the Stakeholders:*
- Is the goal to grade leadership and punish the guilty or to achieve organizational improvements?
  - •Will the unique conditions affecting different locations be recognized?
- If it's not relevant to my location, why include it?
- Why measure things we don't control?

• Can potentially negative elements be omitted from data collection until they improve?

### *From the "Connecting to the Customer" Team:*

• What experience do the various local facilities management teams have with collection and use of customer data and continuous improvement processes?

• Is there any common level of understanding and knowledge about quality improvement processes driven by customer feedback and measurement?

The team implemented the survey and reporting capability using the idTools web-based software system. The reports and data analysis were carefully designed for simplicity and ease of use by both the team and location managers, who were also required to access the reports and use the information. The team decided that data would be collected quarterly from a random sample consisting of 25 percent of the entire population of North American employees at each cycle. Thus after each complete year of collecting data, the entire population would have been sampled.

The survey also collects open-ended comments after each section of the survey, which the Report-User can choose to view or hide. Reports can be generated by any combination of Location, Business Unit, Job Level, and Time Spent in the Office, which are selected through pull down menus at the top of the report.

***Statistical Process Control (SPC).*** Once the data was collected, Statistical Process Control (SPC) can be used to analyze the process variation measures. A primary tool of SPC is the Control Chart, which permits identification of the sources of process variation (discussed in detail in the following section). Tools used by the F&S team include spreadsheet templates, bar charts, and other graphics to track and compare data collected from locations. These tools have the advantage of easy access and simplicity of use for location managers.

Using SPC tools, the variability in a process over time (such as different aspects of the quality of the work environment) can be monitored by collecting data. The team used simple tools to interpret the data, including:

• Bar charts to show mean score for each location versus the U.S. mean for each survey question

• Graphs to show the range and breakpoints of scores for all the survey elements at a U.S. level (to define what constitutes a high and a low score - grading on a curve)

• Rankings of location scores for each category to identify best and worst conditions/practices

• Scores over multiple quarters for each location and for each category to identify trends. This information yields quarterly reports on "blue ribbon" (high performing) areas; identifies performance trends, areas needing improvement and examples of locations that have successfully engineered measurable improvements.

Control Charts track performance measures and reveal whether adjustments need to be made to the process (or to the work environment) to reduce variability in the measures and bring the process back "into control" if needed. Control Charts also determine if process improvement efforts (such as with facilities services or workplace design) are having the desired effects (see Figures 4.8 and 4.9).

Two types of variability can occur in a process (or work environment) that is being tracked, common cause variation and special cause variation (Eckes, 2001). Common cause variation is simply the normal variation about the statistical mean that occurs when data is collected at many points over time. Special cause variation comes from outside events or fundamental problems with the design of the system (such as facility services processes) in terms of meeting employee needs.

The team's goal was twofold: to reduce special cause variation and increase scores on the workspace and service evaluations. In the case of work environment evaluation, special cause variation would be due to lack of fit between employee needs and the design or management of the space in support of those needs. Figure 4.8 shows the results of the project on one measure, "Office Space Supports Collaboration."

The X Bar/R Bar Chart is a powerful tool that shows process variability and trends, and permits interpretation of data so changes can be made to the workplace. Figure 4.9 shows the Grand Mean (mean of the means) of the samples at each observation point over time. The R Chart (sample range, lower chart, Figure 4.8) shows the variability in the scores from the sample at each observation point. In both charts, the green centerline shows the Grand Mean (average of the averages), about which the scores are plotted. The system calculates boundary lines 3 sigma limits above and below the middle centerline. These charts can then be used to determine if the measure is within Six-Sigma limits.

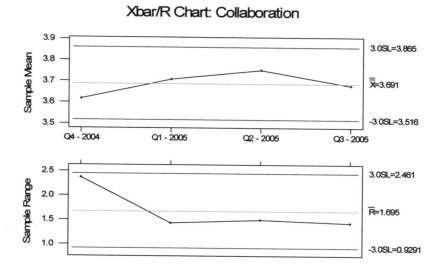

**Figure 4.8** Xbar/R Chart showing "Collaboration" scores (Author)

*Root Cause Analysis.* The information within Control Charts and the other tools used by the team help us conduct a "Root Cause Analysis" of any problems. The first data point (Q4 2004) shows a below-average score on the measure "Office Space Supports Collaboration" (upper part of Figure 4.8). The variance (range) in this sample is also relatively high, although not 'out of control' (crossing the upper sigma limit line) (see lower Chart, Figure 4.8). Thus, the Q4 score is low and highly variable, indicating a weakness in the process for the delivery of workspaces that support collaboration in field offices around the country. However, Figure 4.8 shows that scores in the two subsequent business quarters improved in terms of performance on the measure of collaboration. In addition, the range or variability of the underlying scores in Q1, Q2, and Q3 of 2005 was greatly reduced and stabilized.

A Control Chart was created for a measure of "F&S Service Quality" as determined by their internal customers (see Figure 4.9). Over the four business quarters measured, customer satisfaction with this aspect of service quality remain high (average score = 4.198) (see upper Figure 4.9), and over time the variability underlying these scores was consistently reduced (see lower Figure 4.9), meaning that the process variation within the delivery of F&S services is reduced over time, which is the key to achieving and maintaining process excellence.

3. Improve Phase

In the Improve Phase, a Root Cause Analysis was performed in which the possible causes for the performance gaps between locations related to the Collaborative space measure, were generated by the team along with selected location managers, and a list of the 'vital few' causes were selected. The goal was to reduce problems with workplace design, workplace management issues, or customer perceptions.

Feedback was provided to locations with especially low scores so that processes and workplace design features could be modified to improve scores, and as importantly, reduce the variation in the scores, and thus the variation in the underlying process for delivering collaborative spaces.

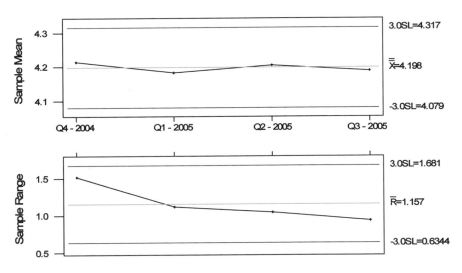

Figure 4.9 Xbar/R Chart showing F&S Service Quality scores (Author)

4. Control Phase

In the DMAIC approach, we attempt to "hold the gains" made through the project, by use of a Control Phase. For the organization, the key components of this phase were the implementation of a monitoring plan, creation of a response plan, and sharing the "lessons learned" with the organization on a continuing basis.

*Monitoring Plan.* As part of the monitoring plan, the team used the idTools survey and reporting system as part of a permanent feedback loop to gather information on service performance. However, there were concerns that some locations might be slow to embrace the use of customer data. To ensure use of this data, the team implemented required procedures for all locations. These required procedures include:

- Each location must access the idTools reports on-line, and enter their location data from the idTools reports into a spreadsheet template that displays the individual location scores versus the U.S. mean
- Each location must share data with customers and facilities staff and prepare and submit an action plan each quarter based on location survey results. To inform the action plan, customer interviews or focus groups may be consulted.

*Response Plan.* The Response plan is a checklist or process by which the team reacts to the analysis provided by the Control Charts when a problem is revealed in one of the performance metrics. In this case, the Response plan is a simple set of steps (the Root Cause Analysis) that allows the team to identify and rank any problems, and brainstorm design responses as required.

### Sharing the Information.

- The "Connecting to the Customer" team prepares a quarterly analysis of data for discussion in a virtual meeting with all locations and capability leads. Tracking the data over time helps this team identify "best practices" that support high performing elements.
- The idTools application allows information to be shared and accessed by all.
- The "Connecting to the Customer" team established a "Round table" with below-manager level employees to conduct interviews, identify best practices, and make recommendations to facilities leadership. The focus on line staff is a deliberate strategy to take advantage of the knowledge and impact of those employees closest to the customer. Survey results provide the basis for identifying areas needing improvement and strength to select locations and customers for in-depth interviews.

## 5. Summary

The "Connecting to the Customer" team wanted an in-house capability that would put them in control of the evaluation process and give them the opportunity for continuous improvement of workplace and services. The team selected the idTools system, an enterprise survey data analysis application for improving the quality of facility management services. The team also created other in-house tools to analyze and display the data and share the information and knowledge gained in an effective manner throughout the organization. Lately the F&S team has begun tracking this data using Six-Sigma quality tools, such as Control Charts. The team couches this quality process as part of a more comprehensive Six-Sigma approach.

6. The Process: Lessons Learned

Communication, involvement, and education about program scope and benefits to location managers is essential to achieve acceptance and imbed the DMAIC process in day to day operations. For some, the process of tracking and using customer ratings to drive continuous improvement may not be familiar or easily understood.

- Measuring performance can be intimidating. It is less threatening when leadership maintains a focus on the constructive aspects of this process.
- Trends on scores over time is more important than the absolute value of an individual score.
- Data must be used to effect change in the organization. If you are not going to use the information, don't collect it.
- In addition to customer satisfaction, the process must address cost-benefit issues and help define what is essential versus "nice to have".
- Customer satisfaction measures should fit into a higher-level framework of a balanced scorecard for the entire organization.
- Survey questions must be carefully designed and worded to yield usable data.
- Customer input, not personal opinion, should define what elements are important and included in survey content.
- Independent access to the web based survey reports is important to assure data is accessible to all locations. This encourages a pro-active approach at the location level and helps create "ownership" of the data and the process, and pride of ownership in the outcomes. Local site managers who must use the data will gain the opportunity to identify successful business practices. This will make improvements easier, faster and less threatening in their own local operations.
- The survey process, when fully integrated into operations, can be used as one component among several to set performance goals and evaluate progress.

**PART III: CASE STUDIES**
**FACILITY/BUILDING, GROUP AND INDIVIDUAL SPACE**

# CHAPTER 5

## Case Studies: Facility and the Organization

### I. INTRODUCTION

This chapter describes five Case Studies in which large scale implementations of flexible workspaces were conducted. The organizations are from the consulting, financial services, and manufacturing industries. The primary population is professional workers, but in one Case Study we examined Call Center agents.

In several of the cases, the change was in part driven by a desire to consolidate space and reduce real estate costs. However, the organizations making these changes also intended to enhance the flexibility of the individual spaces and the overall flexibility of the facility (and thus environmental control). It was thought that this approach would enhance the performance of entire groups and departments.

To test the success of these workspaces, we used the Workplace Performance Measurement Model (described in Chapter 4) to create key metrics to assess changes in behaviors, performance, and business process efficiency. We used this approach to relate any changes we detected in the performance metrics directly back to design features of the workspace.

*Case Study 1.* In the first Case Study we examined the effects of Environmental Control at the group and business unit levels of a professional services organization. The company consolidated employees from four buildings into one, using new furniture and interior design concepts intended to optimize the flexibility of the interior space in supporting individual work and group collaboration. The experimental workplace has fewer workstations than people and a variety of different types of spaces for collaborative work. This was combined with ergonomic training geared to improve ergonomic awareness and communicate new "rules of behavior" for using this space. We hypothesized that the flexible design concepts and ergonomic training would increase human and business performance, and reduce health risks.

We examined three groups of employees, for a total of 1135 participants, at three points in time. These employees were professional and internal support job types. One group served as a "Control" group (no changes in workspace and no training); one Experimental group moved into the new space; and another Experimental group moved into the new space and also received ergonomic training. Data was gathered from all three groups once before, and twice after, the move to the new space and training sessions. The key measures for this study included: musculoskeletal discomfort, job control, collaboration, and quantitative measures of business process efficiency (time and cost of processes). We found improvements on each of the key measures for employees who moved to the experimental workspace, and even greater improvements for those who moved to the new workspace and received training.

**Case Study 2.** This study was undertaken to assess the effects of a move to a more open (and cost effective) work environment on the behavior and performance of 1000 participating Call Center Agents within a "Financial Services Company." Job responsibilities for this type of work have dramatically expanded in the last 10 years. What used to be an individual "help desk" job may now include selling, problem solving, and other interactions that can benefit from collaborating with peers. In this 12 month study, data was collected at four points in time from an Experimental group (three different groups of employees who moved to the new facility) and Control group (employees doing equivalent work, who did not move).

The key measures for this study included: job control, communication, collaboration, sense of community, retention rates (voluntary separations), and three aspects of agent performance, ACD (number of calls handled), AHT (average handle time per call), and ACW (time to complete work after call has ended). We found improvements in job control and collaboration for employees who moved to the new space. We found no changes in ACD or ACW but we did see a significant increase in AHT (average handle time) for calls for agents who received the new space, a positive finding that relates to increased collaboration and quality of service. We also identified the workspace design features that predict each of these agent performance measures. Finally, we found that employees who moved to the new space had 2 percent lower rate of voluntary separations than did employees in the control group, over the 12 month period of the study.

**Case Study 3.** This Case Study discusses the results of a research project that was conducted as part of a building consolidation project for the Headquarters campus of a leading company in the overnight shipping industry. The 720 participants in this research were part of a consolidation from four widely separated buildings into one existing location. As part of the design process, adjacencies between departments and work groups were carefully optimized in the new space. Most of the furnishings from the original locations was reused in the new space, although a limited number of people received new furnishings. The professional employees worked within 18 departments such as HR, Finance, Legal, and other groups typically found within a corporate headquarters.

The purpose of the project was to measure changes in employee behavior, stress, health, job satisfaction, and business process efficiency before and after the consolidation. We employed three methods of assessment, including: a questionnaire given to all employees, managers' ratings of the quality of output of each department (prior to and after the move), and a business process analysis (BPA). In addition to pre and post comparisons, we created regression models that show the workplace design features that predict behaviors and work process improvements. We also created a model that illustrates the financial impact of increased efficiency of business process.

Participant responses were compared on three treatment conditions ("before change," "consolidation only," and "consolidation and new furniture"). Data was collected at one point prior to the move, and at two points after the move.

For workstation evaluations, in the "consolidation only" condition we found consistent improvements in assessments of quality of lighting, storage, workstation layout, and comfort. We found some, but less consistent improvements, for the "consolidation and new furniture" condition.

For evaluations of workspace support for group work, we found that employees who received the consolidation/redesign only, and the consolidation coupled with new furniture, reported significantly greater support for meetings, availability of spaces for training, reduced travel time between airport and workstation, reduced travel time between home and work, and reduced travel time between workstation and all other departments.

We analyzed the impact of consolidation and new furnishings on various aspects of group work process, including: communication between departments, departmental collaboration, departmental responsiveness, face to face collaboration, and other behaviors.

We found no difference in frequency of communication between departments on any of the comparisons. Employees who received the consolidation/redesign only, and the consolidation coupled with new furniture, reported significantly increased workflow and collaboration between departments. We found no difference in self-assessed quality of group process on any of the comparisons. We found an improvement in the quality and accuracy of team reports when compared to employees before any changes to the work environment. We found a significant increase in departmental responsiveness (speed and accuracy of response). We found positive impacts of consolidation/redesign and new furnishings on various aspects of employee health, including back pain, discomfort in hands and arms, and general health. We found no change in self-reported psychological stress levels.

We measured "Business Case Approval Time" by conducting a business process analysis (BPA) of this process, and collecting data on the time and cost of conducting this process from key management leads in four departments. With this analysis, we assessed the amount of time (in days) it took to have a business case approved, before and after the consolidation. We found a 32 percent reduction in time and costs for one process we measured. The annual cost savings in labor dollars for this one process was US$120,000 per year.

We created regression models to determine what, if any, workplace design features predict outcomes, including: sense of privacy, collaboration and workflow, group effectiveness, and speed of Business Case Approvals process.

We found that five variables predicted privacy, including: ability to handle confidential materials in the workstation, level of interruptions, ability to have confidential conversations in the workstation, interior layout of workstation, work surface size, ease of organizing work materials, and storage.

We found that four variables predicted collaboration, including: overall lighting in the workstation, travel time between the workstation and all other departments, amount of interruptions experienced within the workstation, and interior layout of the workstation. We found that collaboration predicted any aspect of group effectiveness.

We found that five variables predicted a large and significant amount of the variance in speed of the approval process, including: adequate spaces for meetings between groups, interior layout of the workstation (work materials close at hand, floor area, arrangement of equipment and furnishings, work surface size, ease of organizing work materials), level of visual and noise distractions, travel time between workstation and all other departments, and workstation has adequate space to support collaboration with another person.

*Case Study 4.* This study assesses the effects of a move due to a building consolidation, and implementation of a more open environment/hoteling model on the behaviors, perceptions, and business process efficiency of individuals and groups for 750 professional employees of a manufacturing company based in the Midwest. In this study, we collected survey and business process efficiency metrics from 750 employees comprising 22 departments at two points in time from an Experimental group (those who relocated to the new facility) and from a Control group with an equivalent number of employees who did not move.

Key measures for this study were collected for individuals, groups/teams, and departments. We also collected business process metrics that cut across departments and groups.

Individual measures included an assessment of quality of workspace features (storage, lighting, layout, comfort, privacy, sense of community, etc.) perceptions of job control, and behaviors such as collaboration and communication. Measures at the Departmental level included assessments of quality of departmental output (responsiveness and quality of deliverables) by all other departments. Measures of business process efficiency were taken on three routinely occurring business processes within three departments, in which detailed time and cost calculations were made.

The analyses showed that there were either no changes (such as in job satisfaction) or actual decrements on some individual assessments of workstation features (such as storage, layout) and related behaviors (privacy, communication) at the individual level. At the group and departmental level, we found significant improvements in the quality of departmental outputs (responsiveness and quality of work) and on assessments of degree to which spaces support collaboration of groups, sense of belonging, departmental identity, potential of space to attract new employees, and communication of corporate identity. The data we collected on three ongoing business processes revealed that one of the three processes showed significant time reduction, with no change in the other two processes.

We then created a series of regression models to determine what features of the work environment might be predictors of the perceptions and behaviors at the individual, group and departmental levels, and of the business process metrics.

We found that job satisfaction was predicted by a combination of five variables related to communication, control over the job, and work environment design features such as storage and layout of space in workstation.

Job control is predicted by four variables, including job type, interruptions, communication between individuals, and space support for individual work.

The quality of group process was predicted by four variables related to aspects of communication and work environment design; including spaces to support collaboration.

We found that quality of the departmental work products is predicted by two variables: availability of collaborative space, and effective communication from management to employees.

We found a 7.5 percent decrease in cycle time for the New Product Specification process that contributes $375 of cost savings to the business each time the process occurs. Based on the number of process occurrences, $4,000 of cost savings are realized annually for just this one process.

We then developed a regression model to determine the predictors of process cycle time for the New Product Specification Process. We found six predictor variables, including; availability of collaborative space, lighting, quality of group process, design of interior space supports shift from individual to collaborative work, amount of time spent in unassigned workspace, and quality of storage in the workstation.

We then plotted process cycle time (for the New Product Specification Process) against the independent assessment of the quality of departmental output for the eMarketing department (which is responsible for the New

Product Specification Process). We found that process cycle time decreased after the move to the Marketplace, and that at the same time, departmental work product quality increased.

These findings would be exciting if we had only found that design features reduced the cost of this process. What this research has shown is that, due to improved workplace design features, process costs have been reduced, and in addition, the quality of the output (beyond the time to produce the product) has been increased. This is a powerful example of a classic definition of productivity. Quality of output rose while the time (cost) to deliver the output was reduced.

***Case Study 5.*** This study assesses the differences between two call centers of a telecommunications company. The call centers used two different models of workspace furnishings, layout, and quality of the architectural space itself. The site in Illinois was a "Class A" building space with high quality finishes and furnishings, with traditional low walled cubicles for agents. The Iowa facility was a converted "big box" retail store and used furnishings that provided a comparatively more open work space. Other than the workspaces, the technology, jobs, and management practices were equivalent at the two locations.

We collected a comprehensive range of subjective survey measures, and objective business and human resource metrics -- all data other than the survey data was independently collected as part of the organizations' ongoing business practices. The data we collected included: a Workspace/Behavioral Assessment (Survey), Automatic Call Distributor (ACD) Performance data, Customer Satisfaction Scores on agent performance, agent Job Satisfaction scores, Claims Data (costs), and Lost Work Days at each site.

The Iowa facility generally outperformed the Illinois facility on employees' evaluations of the supportiveness of the workspaces in their work, had better ACD scores, better Customer Satisfaction scores, lower claims, and fewer lost days. The Illinois facility had better job satisfaction scores. An additional focus of this study was less on a direct comparison, and more on using the data from both locations to create regression models to understand the larger design principles that might predict the various performance outcomes.

We report a few key findings here. We found small but consistent and statistically significant relationships between workspace features and several of the key outcome measures. For instance, as workspace support for collaboration increases, so does percentage of First Call Resolution, a key cost and strategic business driver. As adjust ability of workstation features and job control increased, external measures of Customer Satisfaction with agent performance increased. As agent pain and discomfort decreases, and group cohesion increases, "after call" work time decreased. We report numerous other relationships within the full case study later in this chapter.

The types of metrics we used in these Case Studies are groundbreaking in the sense that they include measures such as Call Center agent performance measures (number and time of calls), and business process metrics (time and cost to conduct a process) for professional workers. Throughout this chapter we use these data to develop statistical models that explicitly show the relationship between these quantitative performance metrics and specific features of the designed workspace.

## II. CASE STUDY: EFFECTS OF FLEXIBLE WORK ENVIRONMENT DESIGN AND TRAINING ON PROFESSIONAL EMPLOYEE PERFORMANCE AND BUSINESS PROCESS EFFICIENCY

The purpose of this study was to examine the effects of Environmental Control at the group and business unit levels of a professional services organization. Environmental Control was "designed in" to the workplace through a flexible workspace model that included training of managers and employees on use of this space.

This study employed performance metrics such as: musculoskeletal discomfort, job control, collaboration, and quantitative measures of business process efficiency (time and cost of processes). The 1135 employees who participated in this project were professional and internal support job types, with all or most work conducted at this location.

### A. Workplace Issues

The central issues that the organization was attempting to improve through design were: enhancing collaboration, comfort, job control, and sense of community.

### B. The Workspaces

The organization sought to rethink the traditional use of office space at one location by organizing the use of space by function and emphasizing flexibility of space use and assignment of work space. The existing workspaces provided one furniture workstation for each employee. These workspaces were uniform in nature and were progressively larger depending on job level (see Figure 5.1). Workstations provided seated privacy and were designed to support individual, "heads down" work.

**Figure 5.1** Space prior to renovation (Author)

There were only a limited number of spaces to support group work and those were mostly used for larger formal meetings.

The company moved employees from four buildings into one, using new furniture and interior design concepts intended to optimize the flexibility of the interior space in supporting individual work and group collaboration. The experimental workplace has fewer workstations than people and a variety of different types of spaces for collaborative work.

The scope of this re-design project involved 700 workspaces in a building area of 125,000 square feet. In this new workplace model, the numbers of employees with assigned workspaces were greatly reduced, while a large percentage of the workspaces are now "unassigned." In this new model, the use of workspace size and location to indicate rank or status was purposefully diminished. In fact, the most desirable spaces might be unassigned spaces by the window wall (see Figure 5.2). The design emphasis of the new approach was on supporting mobility of employees, flexibility of space use, and ease with which employees can shift from individual work to group work, and back again. Figure 5.3 shows a typical small meeting space found within the facility that employees can use as required during the course of the day.

Figure 5.2 Unassigned individual workspace (Author)

## C. Training

The training consisted of a 45-minute presentation to employees and managers on basic ergonomic principles, and information pertaining to new "rules of behavior" that apply to working in the new space. Employees were trained in groups of 20. The emphasis of the latter portion of the training session was to ensure that managers and employees felt they had "permission" to work in locations (such as the many small alternative meeting spaces, cafeteria, lounge, etc.) other than in their assigned or unassigned workspace. The overall intent of the training was to ensure that people took full advantage of the flexibility and control that the new space would give to their work process and patterns of use of the space, and allow them to use the space to best match their ongoing individual and group work process demands.

## D. Study Hypotheses

Figure 5.4 provides a graphic representation of the 2 x 2 factor design used in this study. There were two levels of workspace flexibility (low versus high), with the original workspaces representing the traditional "low" amount of flexibility and support for mobility and group work. These levels are shown in Figure 5.4 by the vertical axis. There were two levels of Training in this study: the "Yes" condition indicates participants who received training on how to use the new space, and the "No" condition indicates employees who did not receive training. Due to changes over the course of the study, this design ultimately did not include a condition in which employees in the "low" flexibility condition received training, so it was not possible to determine the effects of training alone on the outcomes.

### 1. Hypothesis 1

Key desired behaviors, perceptions and performance measures (Job Control, collaboration, communication, sense of community, and business process efficiency) would be most improved in the "high" flexibility with training Condition (see upper left quadrant, Figure 5.4).

### 2. Hypothesis 2

The greatest reduction in musculoskeletal discomfort will result from the combination of "high" workspace flexibility and ergonomic training (see upper left quadrant, Figure 5.4).

**Figure 5.3** Unassigned individual workspace (Author)

### E. Methods

#### 1. Survey

Using a 72-item questionnaire, we analyzed variables assessing attitudes and behaviors of employees in their workplaces. The response scales for the question items used a 5-point Likert-type scale (agree-disagree) or (satisfied-dissatisfied) as required for the question.

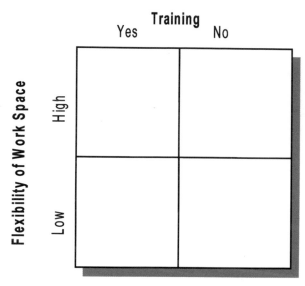

**Figure 5.4** Illustration of study hypotheses (Author)

2. Research Design

These variables were measured at three points in our project (once before, and twice after, two of the groups moved into a newly designed workspace), and the same questions were asked of three groups (one that did not move, one that moved into a new space, and one that moved into a new space and received training about ergonomic and other aspects of their new workspace). The three sets of observations were spaced 5 months apart from each other.

3. Business Process Analysis

We also created detailed process maps of four internal business processes that occurred within the Experimental group, and an additional four identical processes occurring within the Control group. The process maps included metrics on the time and cost to accomplish each step of the process, and permitted us to calculate actual process costs using annualized compensation data. The process maps and time metrics data were collected through interviews and direct observations of employees conducting those processes.

4. Analysis Plan

We used repeated measures analyses of variance (MANOVA) to look for significant changes "within group" in the variables across those three observations. These analyses were also conducted by treatment group (experimental versus control). In the Results section, the means for the variables are reported, broken down by observation and treatment group, together with an interpretation about the statistical significance of any changes, supported by the analyses of variance. We also used line graphs to illustrate the different patterns of change on key measures, among the three treatment groups.

The repeated measures analyses of variance assess the significance of the effects in several ways. For the purposes of this study, we were most interested in whether there were interactions between the experimental and control groups. In other words, were the changes over time different between the control and experimental groups?

*The central question is whether or not the experimental group improved over time in a way not also seen in the control group.* This should be clearer when looking at the line graphs. The statistical significance of those different patterns is assessed by the interaction term of the repeated measures analysis of variance. Genuine treatment effects will be reflected in improvements in the two treatment groups combined with less or no improvements in the control group (or actual declines), with a significant interaction term.

## III. RESULTS

### A. Workspace Evaluation

We found that the Experimental group was more satisfied with 10 aspects of the workspace design, after the move to the new space. These variables showed significant increased satisfaction in the Experimental treatment group on the variables shown in Table 5.1.

At the same time, though, the treatment groups experienced increased distractions. This finding is congruent with the results of other studies in which a move to more open, functionally effective space is paired with increased distractions.

## B. Impact of Workplace Design and Training on Job Control

Research from large-scale longitudinal studies show that those whose jobs allowed them greater decision-making control in doing their work had health that remained largely unchanged over a four-year study (Karasek and Theorell, 1990).

| Table 5.1 Increased Satisfaction with Workspace Features for Experimental Group |
|---|
| • Comfort<br>• Arrangement of furnishings and equipment (work surface, storage, chair, computer, etc.) in the workspace<br>• Size of the work surfaces (desks, tables) in the workspace is appropriate for work needs<br>• Lighting in workspace (overhead, task)<br>• Amount of natural light at the workspace<br>• Acoustic privacy<br>• Space supports face-to-face confidential conversations<br>• Degree to which workspace (whether assigned or unassigned) supports work needs<br>• Adjustability of workspace to fit needs<br>• Availability of different workspace settings to accomplish individual work |

Those who described their work as being low in job control, high in job demand, or who reported low levels of work-related social support had health that began low and worsened over the course of the study. In another study, health issues considered included broad quality of life issues such as the ability to carry out daily household chores, social support, and general mental health. Feelings of job control are clearly important in terms of employees' long-term health as well as minimizing sick days in the short-term. The fact that Job Control has such important effects is why it is included within this study.

The Job Control factor used in this study included four components: participants' sense of control over (their) work tasks, control over their schedule, control over selection and use of their work tools, and control over selecting their location of work. Higher scores indicated greater sense of job control.

Job control is a reliable predictor of employee health and reduction of risks of coronary heart disease (CHD). All employees who moved into the new work environment reported a small (3 percent) but statistically significant increase in job control. Those who also received training reported a 13 percent increase. For all of the individual variables making up the composite measure of Job Control, there was evidence of the effectiveness of the treatments, and at the same time, for each variable the control group showed a decline over time.

Thus, on this powerful construct of control, we found consistent positive benefits of the new flexible workspaces (and in some cases even stronger effects for the workspace + training factor) on sense of job control. Because a positive sense of job control has been consistently linked to health issues, this is an important finding for the development and use of these types of spaces for professional workers.

## C. Impact of Workplace Design and Training on Comfort

The variables assessing Comfort measure the work-related discomfort that participants' report experiencing in various parts of their bodies. Participants were asked to "rate the pain or discomfort for each body part." A '0' indicated no pain, while 1, 2, 3, 4, and 5 represented "Just Noticeable," "Some Discomfort," "Definitely Uncomfortable," "Very Uncomfortable," and "Extremely Uncomfortable." We asked participants to assess discomfort on: neck, shoulders, upper and lower back, arms, hands, fingers, and legs. Using this coding scheme, lower scores represent greater comfort.

We found a significant pattern of reduced pain in: neck, upper back, and lower back for participants who moved to the new workspace, and greater effects for those who moved to the new workplace and also received the training.

We also asked participants to indicate whether they had experienced work-related discomfort in the previous 4 weeks. The results were reported as a yes/no item. Table 5.2 shows the percentage of respondents indicating that they had experienced work-related discomfort in the previous 4 week period.

The results show that while the Control group reported an overall gain over time in the percentage of participants reporting work-related pain, the group who moved to the new workspaces saw a significant decline in work-related discomfort (see Table 5.2). For the treatment group that received both the new workspace and training, the decline in discomfort was even greater, declining from over 76 percent to around 30 percent over the course of the study (see Table 5.2).

| Table 5.2 Percentage of Sample Reporting Work-Related Discomfort within Previous 4 Weeks | | | |
|---|---|---|---|
| Treatment Group | Observation 1 (Pre-Move) | Observation 2 | Observation 3 |
| Control | 43.9% | 35.0% | 51.2% |
| Experimental Workspace | 35.5% | 31.3% | 30.3% |
| Experimental Workspace + Training | 75.5% | 52.9% | 25.4% |

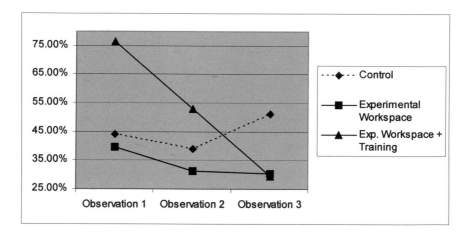

**Figure 5.5** Work-related discomfort by treatment group (Author)

Figure 5.5 shows the results in graphic format. The most obvious result in Figure 5.5 is the dramatic drop in pain or discomfort reported by the members of the treatment group that received both the new workspace and the Training. "Work-related discomfort" decreased significantly (27 percent) for employees who moved to the new space, and decreased by 46 percent for those who moved and also received training (see Table 5.2). Both these findings are statistically significant.

## D. Impact of Workplace Design and Training on Collaboration

We assessed collaboration from the following individual question items:
• design of the various spaces in this office provides adequate support for collaboration;
  • appropriate space is available to me to collaborate when I need it;
  • appropriate technology tools (display, video/web conference capabilities;
  • power/data connection, etc.) are available in the meeting spaces;
  • the workspace lets me quickly shift from individual work to collaboration with others.
We created an index of collaboration using these items, and analyzed the change over time for all treatment groups. The results of these analyses are shown in Figure 5.6. All four of the collaboration variables showed evidence of improved satisfaction due to the treatments.

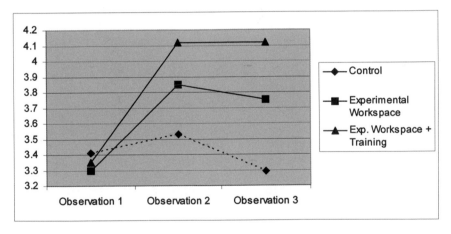

**Figure 5.6** Collaboration by treatment group (Author)

Figure 5.6 shows that at Observation 1 (prior to the move or training) collaboration level was similar for all groups. Over time, collaboration for the Control Group remained stable and slightly declined. Collaboration for the group that received the new space increased significantly after the move, then stabilized over time. Collaboration for the group that received the new space and training increased significantly after the move, and also stabilized at the final Observation (see Figure 5.6). Participants who received the new workspace and training had the greatest increase in collaboration, followed by those who received only the new workspace.

## E. Impact of Workplace Design and Training on Communication

We assessed communication from the following individual question items: amount of face-to-face interaction with people outside my work group, amount of face-to-face interaction with members of my work group, ease of access to the employee, and of the employee to coworkers, and overall quality of communication. We created an index of communication using these five items, and analyzed the change over time for all treatment groups. The results of these analyses are shown in Figure 5.7.

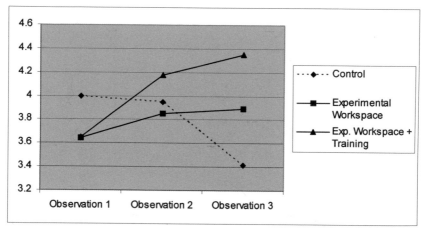

**Figure 5.7** Communication by treatment group (Author)

Figure 5.7 shows that at Observation 1 (prior to the move or training) Communication level was actually higher for the Control group than for the other two treatment groups. Over time, communication for the Control Group remained stable and then slightly declined (see Observation 3, Figure 5.7). Communication for the group that received the new space increased (but not significantly) after the move (Observation 2), and then increased significantly by Observation 3. Communication for the treatment group that received the new space and training increased significantly after the move, and also continued to increase by the final Observation (see Figure 5.7). Participants who received the new workspace and training had the greatest increase in collaboration followed by those who received only the new workspace.

## F. Impact of Workplace Design and Training on: Community

We assessed the degree to which the overall workspace conveys a "sense of community" from the following individual question items, workplace supports: sense of belonging, identity with work group, sense of ownership in the organization, conveys appropriate image to employees, customers and investors, and attracts talented people. We created an index of Community using these six items, and analyzed the change over time for all treatment groups. The results of these analyses are shown in Figure 5.8.

Figure 5.8 shows that at Observation 1 (prior to the move or training) sense of Community was actually higher for the Control group than for the other two treatment groups. Over time, sense of Community for the Control Group remained stable and then significantly declined (see Observation 3, Figure 5.8). Sense of Community for the group that received the new space increased after the move (Observation 2), but was not significantly greater than the Control Group. However, by Observation 3 employees in the Experimental workspace reported significantly greater sense of Community than did the Control Group (see Figure 5.8). Sense of Community for the treatment group that received the new space and training increased significantly after the move, and also continued to increase by the final Observation (see Figure 5.8). Participants who received the new workspace and training had a significantly greater sense of Community than the participants who only received the new workspace, both immediately after the move (Observation 2) and later (Observation 3) (see Figure 5.8). Thus, both treatment groups saw significant benefits in terms of enhanced sense of Community, with employees who received training seeing the greatest effects.

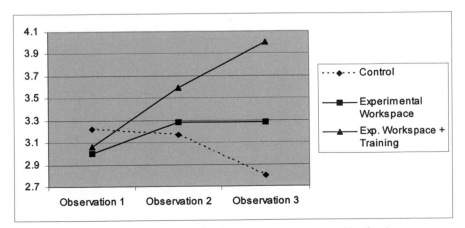

**Figure 5.8** Sense of Community by treatment group (Author)

## G. Impact of Workplace Design and Training: Group Effectiveness

We assessed Work Group Effectiveness from the following individual question items: Problem Solving, Making Decisions, Getting the Work Done, Making Use of Member Skills, Accuracy of Reports, and Developing New Ideas. We created an index of Work Group Effectiveness using these six items, and analyzed the change over time for all treatment groups. The results of these analyses are shown in Figure 5.9.

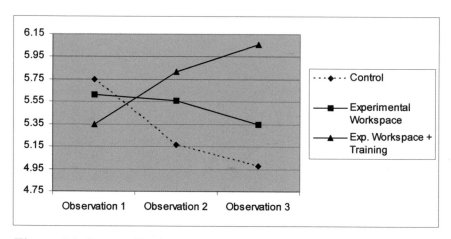

**Figure 5.9** Group effectiveness by treatment group (Author)

Figure 5.9 shows that at Observation 1 (prior to the move or training) Group Effectiveness was actually higher for the Control group than for the other two treatment groups (but not significantly different). Over time, Group Effectiveness for the Control group significantly declined (see Observation 3, Figure 5.9). Group Effectiveness for the group that received the new space did not change between the pre and post move observations (see Figure 5.9) but declined significantly at Observation 3. However, employees who received the Experimental space still had significantly higher Group Effectiveness scores than the Control group on the two post-move measures. When compared to the Control group, Group Effectiveness for the treatment group that received the new space and training increased significantly after the move, and also continued to increase by the final Observation (see Figure 5.9). Participants who received the new workspace and training had significantly greater Group Effectiveness than the participants who only received the new workspace, both immediately after the move (Observation 2) and later (Observation 3) (see Figure 5.9). Thus, both treatment groups saw significant benefits in terms of enhanced Group Effectiveness, with employees who received training seeing the greatest effects.

## H. Impact of Workplace Design and Training on Process Efficiency

We employed Business Process Analysis (BPA) methods to track the time and cost related to eight ongoing internal business processes. These analyses were conducted for both control and experimental groups. The processes include:

- The 'Quarterly Financial Reporting' process was unchanged in both time and quality. This was as expected, since the group responsible for this function was within the Control group and did not receive a new work environment or training.

- The time to complete the 'Monthly Financial Reporting' process was reduced by 4.28 percent per execution of the process, while quality of output remained constant at previous levels.

- The time to complete the 'Project Scheduling' process was reduced by 15.15 percent per execution of the process, while quality of output increased.

- The time to complete the 'Performance Review' process was reduced by 4.52 percent per execution of the process, while quality of the output increased.

| Table 5.3 Summary of Reduction in Process Cycle Time | |
|---|---|
| Treatment Group | Reduction in Process Cycle Time[1] |
| Experimental Workspace Only | 5.62% |
| Experimental Workspace and Training | 10.55% |
| [1] Reduction is calculated across all four processes that we measured. Percentage of reduction is pre-treatment process time versus post-treatment process time. | |

Table 5.3 shows a summary of the overall percentage change across the 4 business processes that we analyzed. Overall, the group that received the new workspace saw a 5.5% reduction in process cycle time and cost. The group that received the new workspace and training reduced process cycle time by 10.55%. There was no change in process time for the four similar processes we tracked within the Control Group.

Further regression analyses of the data showed that the flexible workspace and training (when treated as factors in a multiple regression equation) were directly responsible for process cycle time reductions. This finding is critically important, since it illustrates the direct effect of workplace design and training on quantitative measures of organizational output – a measure of productivity.

Our analyses also revealed that the relative size of the effects of the New Workspace and Training were equivalent in their impact on business process efficiency. Interestingly, these effects are additive; that is, the effects of training add on to process time reductions, on top of the positive effects of the work environment. The two aspects of treatment together accounted for roughly 23 percent of the total variance in the time savings across the four processes.

## IV. CONCLUSIONS

The strategic investments made in the physical work environment are really investments in people, and it is people who will constitute an organization's competitive advantage in the global economy. This company successfully created a design that supports appropriate departmental and group adjacencies and meeting and workspaces to support collaboration and improve the overall quality of work products. The results of this study indicate that the investment made has resulted in significant improvements in group work practices and process efficiency.

# V. CASE STUDY: EFFECTS OF OPEN FLEXIBLE WORK ENVIRONMENTS ON CALL CENTER AGENT PERFORMANCE

## A. Background

This study was undertaken to assess the effects of a move to a more open work environment on the behavior and performance of Call Center Agents within a "Financial Services Company" (the name we will use for this Case Study). In this 12 month study, data was collected at four points in time from an Experimental group (three different groups of employees who moved to the new facility) and Control group (employees who had not moved). A total of approximately 1000 employees, primarily Call Center Agents, participated in this study.

## B. The Workspaces

The new call center was designed to be not only cost effective itself, but to promote increased work performance for Call Center Agents. Job responsibilities for this type of work have dramatically expanded in the last 10 years. What used to be an individual "order-taking" job may now include strategic selling, sophisticated problem solving, and other activities that can benefit from a collaborative, team-oriented approach. The organization hoped to design the new workspace in such a way as to promote desired behaviors and perceptions, including: enhanced sense of job control, communication, collaboration, and sense of community. Related to sense of community, the design of the new space was also intended to positively affect staff retention. The workstations themselves would also be designed to promote a high level of individual control over the immediate space, to enhance job control, and control over work flow. The original workstations were fairly rigid in design, lacking flexibility and adjustability (see Figure 5.10). The workstations were not designed to support informal collaboration. The overall layout of the workstations on the floorplate was in long anonymous rows of cubes that reinforced the individual nature of the work, having limited spaces for meetings (see Figure 5.11).

The new workstations that participants within the Experimental group moved into are much more open and flexible in their design and have a higher level of adjustability of the features and work tools within them (see Figure 5.12). In addition, the overall layout of the workspaces on the floorplate is "organic" in nature, reflecting the desire to move away from the machine metaphor of design and to a biological metaphor that more accurately reflects the underlying systems and goals of the organization (see

Figure 5.13). This organic layout was also employed to facilitate collaboration and efficiency of workflow, and to enhance a sense of community within the space.

## 1. Study Hypotheses

We predicted that the new, more flexible and open work environment design solution would lead to increased communication and collaboration; that it would provide an increased sense of belonging and job control; and provide and an enhanced quality of group work. We also predicted that the new work environment would positively impact Human Resource issues such as retention and Call Center Agent performance measures.

**Figure 5.10** Typical Call Center agent workstation prior to redesign (Author)

## 2. Data Collection

Three types of data were gathered in this study, including survey data concerning employees' behaviors and perceptions of the work environment, measures of work performance for Call Center Agents, and measures of voluntary and involuntary separations over the course of this study from a Human Resources database.

## 3. Survey Data

Data on employee behaviors, perceptions, and workspace evaluations was gathered through a survey, which contained question-items that were assembled from pre-existing surveys employed in other published studies, as well as new items developed specifically for this study.

**Figure 5.11** Layout of workstations prior to redesign (Author)

The survey gathered employee evaluations of the workspace, including: storage, workstation size, work surface size, comfort, lighting, seating and workstation adjustability; and ability to handle confidential materials. The survey asked employees about their behaviors, including: communication, support for and ease of collaboration, group work process, and job control. The survey also assessed employee perceptions of: noise, privacy, and how workplace design communicates corporate culture.

**Figure 5.12** Layout of workstations after redesign (Author)

## 4. Call Center Agent Metrics

Call Center Agent performance measures were collected on a monthly basis over an 18 month period. These measures included:

- ACD: Automatic Call Distributor. The number of incoming calls received by each agent. In this study, measured as the monthly average of the number of incoming calls received per Call Center Agent, the monthly average per operator being the unit of analysis, over the course of the study.
- ACW: After-Call Work. This is defined as work that is necessitated by and immediately follows an inbound transaction. The agent is unavailable to receive another inbound call while in this mode. Measured as the monthly average of the seconds of time per transaction per Call Center Agent, per month, the monthly average time in seconds per operator being the unit of analysis, over the course of the study.
- AHT: Average Handle Time. The sum of Average Talk Time and Average After-Call Work. Measured as the monthly average of the seconds of time per transaction per Call Center Agent, per month, the monthly average time in seconds per operator being the unit of analysis, over the course of the study.

5. Human Resource Data

Data on voluntary separations among participants was collected during the course of the study. Separations were tracked for the Control group and for each move group on a monthly basis throughout the course of the study.

## B. Research Design

A 2 x 4 factorial design was used in the study. Data were collected at four points in time: once before any interventions, and then approximately 60 days after each of the experimental groups moved to their new work environment. Figure 5.13 illustrates the timing of the survey administration and the moves for the various groups involved in the study.

The two primary groups of participants in the study included an Experimental and a Control group. Within the Experimental group were three groups of employees who moved in succession (Group One, Group Two, Group Three) into the new, open work environment.

Three groups from the same location were used as Control groups (Group One Control, Group Two Control, and Group Three Control). The demographic composition of the Control groups were matched with the Experimental groups in order to have a similar mix of departments, job types, tenure with the company, and age.

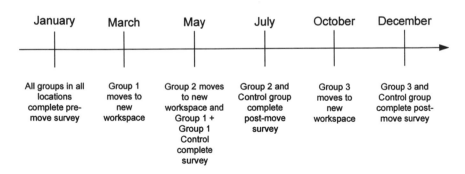

**Figure 5.13** Study Timeline (Author)

1. Advantages of the Experimental Design

The use of the matched sample methodology reduces the possibility of selection bias. That is, the similarity across groups reduces the chance that some kind of inherent differences within the Experimental groups might cause them to respond differently than others to the experimental condition (the new work environment) and therefore limit the generalizability of the results.

Using pre- and post-change surveys allows for an objective measure of the impact of the experimental condition. Each of the three matched groups from the Experimental and Control conditions completed the survey at the beginning of the year before any moves took place in order to get a baseline measure of survey responses. Then each group completed the same survey again following the experimental intervention.

Another benefit of using a pre- and post experimental design with a Control group is that it allows for direct comparisons between the Experimental group's change in responses following the workplace change and the control group's change over that same time. The generalizability of research results is improved by controlling for factors that might unwittingly influence the observed changes: For example, if an organization announces a major new policy during the time between the pre- and post workplace change surveys. That new policy may cause a considerable drop in morale, which could result in individuals providing more negative responses on the post-change survey.

Therefore, looking solely at the Experimental group's responses might cause one to conclude that the new environment led to greater dissatisfaction when in actuality the dissatisfaction was due to the policy change, not the new environment. However, finding that the control group's responses changed in a similarly negative way, despite no work *space* change, would lead to the conclusion that something else may be causing the changes. Comparing changes in responses between the Experimental and Control groups, then, allows for an additional check on the Experimental group's changes.

## 2. Study Participants and Demographics

The three groups (One, Two, and Three) that moved into the new work environment were from the office in the Southwest. Similar groups from a separate office were used as a Control; that is, the Control group employees did not move to a new space and therefore served as a comparison against changes in the responses of the move groups. Most (87 percent) of the study participants were Call Center Agents in both the Experimental and Control conditions. In addition, most of the participants in this study had been with the company for less than three years (60 percent experimental group, 45 percent control group). The average tenure (time employed by the company in years) was 4.5 years for the Experimental group and 4.3 years for the Control group.

The response rate for the pre-move survey was 65 percent for the move groups combined, and 60 percent for the combined Control groups. The Experimental group and their corresponding Control group completed the survey again about two months after the Experimental group moved to their new work environment. Overall, these are reasonable response rates (20 percent is typical for survey research) and reflect employee interest in their workplace and support for this research process.

## 3. Analysis Plan

The analysis plan for the study examined the change in aggregate responses (average responses for all employees in the Experimental group and all employees in the Control group) to the survey questions from before, to after, the move to the new work environment.

Comparing changes across the two groups allowed us to determine whether changes in responses were likely to be related to the new workspace (different amount of change within the move group compared with the Control group) and to use the amount of change in scores from the Control group as a baseline to assess the statistical significance of changes seen in the scores for the Experimental group. The use of a Control group lets us control for changes in scores in the experimental group that might be due to outside events that could affect the organization and employees' perceptions, behaviors, or performance.

In addition, we analyzed the impact of department membership, job type, and tenure with the company as factors potentially affecting the outcome measures. Any differences due to these factors are reported as applicable within this report.

A second component to the analysis involved examining Agent Performance Scores measures and voluntary separations of Agents in all groups from before, to after the move, to determine if these measures were impacted by the new work environment. Over the course of the study, we collected data on the number of employees who were separated each month, by department. We analyzed the data to determine if there was a relationship between use of the new work environment and terminations, or if employee perceptions or behaviors within the new environment were related to terminations.

Correlation analyses were used to identify work environment design features, employee perceptions or other variables that were related to changes in Agent performance metrics data and separations.

## 4. Method of Analyses

This section details the results of the analyses and focuses in particular on the change in aggregate employee responses from before, to after moving to the new work environment for the Experimental group; and change in employee responses in the Control group, over that same time period. Multiple question-item indexes thought to represent key behaviors and perceptions were created, and their reliability tested using Chronbach's alpha. Mean scores were computed for each index for both the pre- and post-move surveys.

These scores were then separated into Experimental (move groups) and Control groups, with an aggregate mean score before and after the move for each. An ANOVA (analysis of the variance between groups) was calculated for each index score, to determine if the experimental and control mean change scores were significantly different from each other. We also conducted ANOVAs to assess differences in responses by department, job type and tenure with the company. In the case of analyses in which these factors were significant, those analyses are described within that section of the discussion.

For the analyses of workstation features, we did not create scale scores but analyzed the data by calculating t-tests between the aggregate means of individual question items that represent various elements of the work environment.

## A. Results

### 1. Workstation Features

In the survey participants responded to a series of questions regarding the features of their workstations and perceptions of their experiences in using the space, including: workstation interior layout, size of work surfaces, amount of light, noise level, conversational privacy, visual privacy, adjustability, storage, access to technology, and chair comfort.

The results of our analyses with the Experimental group show a pattern of enhanced collaboration coupled with reduced privacy and greater perceptions of noise. This is similar to what other companies have found when moving employees to relatively more open workspaces.

***Increases in Satisfaction with Workstation Features.*** Employees in the Experimental group reported a significantly greater increase than the Control group in their satisfaction with the size of their work surfaces ($t = 5.31$, df $= 822$, p $<. 05$); the amount of light at their workstation ($t = 5.10$, df $= 750$, p $<. 05$); their comfort with working with confidential materials at their workstation ($t = 5.18$, df $= 788$, p $< .05$); seating comfort ($t = 1.77$, df $= 773$, p $< .05$); and the ease of adjustment of their chair ($t = 2.55$, df $= 856$, p $< .05$). These results are summarized in Table 5.4.

Table 5.4 Change in Satisfaction with Workstation Features in Experimental Group Compared to Control Group

| Increased Satisfaction | Decreased Satisfaction |
|---|---|
| Workstation size | Noise |
| Working with confidential materials | Conversational privacy |
| Amount of light at workstation | Access to technology tools |
| Seating comfort and adjustability | Workstation adjustability |

Conversely, employees in the Experimental group reported a significantly greater decrease in satisfaction than the Control group on their satisfaction with noise level in their work area ($t = -5.08$, df $= 788$, p $<. 05$); conversational privacy at their workstation ($t = 5.13$, df $= 851$, p $<. 05$); the ability to adjust the workstation to fit their needs without requesting assistance ($t = -6.67$, df $= 788$, p $<. 05$); and having convenient access to needed technology tools ($t = -2.75$, df $=853$, p $<. 05$) (see Table 5.4).

***Employee Perceptions.*** In this section, we created and analyzed two composite variables which included: workspace and technology support for collaboration, and workspace design communicates corporate culture. The definitions for each of these composite variables are provided within each subsection describing the analysis. These results are summarized in Table 5.5.

| Table 5.5 Experimental Group's Change Relative to Control Group in Satisfaction on Behaviors and Perceptions | |
|---|---|
| Behavioral Measure | Change in Experimental Group Relative to Control Group |
| Design Creates Sense of Community | No change overall |
| Communication | No change overall |
| Workspace and Technology Support for, and Ease of, Collaboration | Overall positive increase |
| Job Control | Overall positive increase |
| Group Work Process Quality | No change overall |

***Design Creates Sense of Community.*** We created an index to assess the degree to which the design of the space creates a sense of community in employees. This index of "community" is composed of four question items, including: the design of workspace contributes to sense of belonging to a workgroup; the design of the workspace helps me to feel I am part of The Company; the design of the office space conveys the appropriate image of The Company to employees; and, the design of the office space helps The Company retain talented people. The scale score reliability (post-move) for this index is alpha =.90, an acceptable level of reliability for this composite measure.

The analysis revealed no overall difference in change between the Experimental and Control groups on the index of "community." Scores on this composite variable declined for both experimental and control groups over the course of the study. Thus, while the move to the new space did nothing to enhance sense of community in employees, neither did it negatively impact these perceptions. We also conducted analyses to investigate the effect of department membership, job type, and tenure on "community."

We found no statistically significant effect for department, job type, or tenure on this composite index.

2. Employee Behaviors

In this section, we created and analyzed three composite variables, including: communication, job control, and group work process. The definitions for each of these composite variables are provided within each subsection describing the analysis.

***Communication.*** We created an index to assess "quality of communication," which was composed of three question items, including: ability to access others at work, ease of being accesses, and the overall ease of face-to-face interaction. The scale score reliability (post-move) for this variable is $\alpha$ = .88, an acceptable level of reliability for this composite measure.

We found no significant change in communication for the Experimental group's responses after these employees moved into the new work environment. We also found no effects for department membership, job type, or tenure on this variable.

***Workspace and Technology Support Collaboration.*** We created an index to assess quality of workspace and technology support for collaboration that was composed of three question items, including: adequacy of meeting spaces for collaboration; the availability of appropriate technology in meeting spaces; and, the degree to which the overall workspace allows for quick shifts from individual to collaborative work. The scale score reliability (post-move) for this variable is $\alpha$ =.52, a low but acceptable level of reliability for this composite measure.

The Experimental group reported significantly greater improvement on the collaboration index than did the Control group ($t$ = 6.76, $df$ = 2, $p$ <.05). Thus, employees who moved to the new work environment reported significant improvements in their evaluation of meeting spaces support for collaboration, availability of technology in those spaces to support group work, and enhanced ability to move from individual to collaborative work. We found no effects for department, job type or tenure on this variable.

***Job Control.*** We created an index to assess the degree of job control, that was composed of three question items, including: the ability to choose an appropriate location to do their work; control over the rate at which the employee must work; and control over decisions about how to do their job (job latitude). The scale score reliability (post-move) for this index is $\alpha$ = .76, an acceptable level of reliability for this composite measure.

The analysis revealed that the Experimental group reported a significantly greater increase in Job Control than did the Control group ($F$ = 6.55, $df$ = 2, $p$ <.01). Thus, employees moving into the new work environment reported they had a greater sense of job control than did employees remaining in the existing work environments.

Next, we conducted an analysis to determine the effects of job type on the job control index. We found that Call Center Agents in the Experimental group reported a significantly greater sense of job control than Call Center Agents in the Control group. Thus, the Call Center Agents that moved to the new work environment experienced greater job control when compared to Call Center Agents who remained in the existing original workspaces.

## 3. Group Work Process Quality

We created an index to assess quality of group work process, which was composed of three question items including: participation by group members in group activities; supportiveness of group members to each other; and whether the workgroup had adequate knowledge of how to use technology tools. The scale score reliability (post-move) for this variable is $\alpha = .85$, an acceptable level of reliability for this composite measure.

Overall there was no change between the pre- and post- move conditions on this group work process for either the Control or Experimental groups.

However, when we examined the individual question items we found that after the move, the Experimental group reported significantly greater increase in workgroup knowledge of how to use technology tools, than did the Control group (F = 5.54, df = 1, p<.05). We do not think that the Experimental group had any additional technology training in terms of using workgroup technology tools, but we suggest that perhaps the improved design of group spaces and technology integration into these spaces may have contributed to making the technology easier to use, thus explaining this finding. We found no effects of department membership, tenure, or job type on group work process.

## 4. Workplace Design Variables that Predict: Communication, Collaboration, Job Control, and Community

Multiple regression analyses were used to identify work environment design features and/or employee perceptions that predict four key variables of interest to this study: communication, collaboration, job control, and community. We used the survey responses of all the post-move respondents to develop regression equations for each composite variable. We then calculated regression equations for each composite variable. The results here are shown for Call Center Agents.

***Prediction of Communication.*** Table 5.6 lists the question items that comprise the communication index.

| Table 5.6 Communication Index |
| --- |
| • Ease of accessing co-workers in person. |
| • Ease of accessibility by others. |
| • Satisfaction with amount of face-to-face interaction. |

A regression analysis was conducted for Call Center Agents. For these employees, the equation included five variables, and explained 71 percent of the variation in the communication index. The predictor variables are listed in Table 5.7.

Table 5.7 shows that the variables predicting good communication for

| Table 5.7 Variables predicting Communication for Agents |
|---|
| • Ability to work with confidential materials in workstation. |
| • Seating comfort. |
| • Adequate meeting spaces. |
| • Ability to choose location to do work. |
| • Design of workspace engenders sense of group membership. |

Agents include: ability to handle confidential materials in workspace, comfort, adequate meeting spaces, control over choice of work location, and feelings of belonging to work group.

***Prediction of Collaboration.*** Table 5.8 lists the question items that comprise the collaboration index.

For Agents, we found a regression equation that predicted collaboration

| Table 5.8 Collaboration Index |
|---|
| • Adequate meeting spaces. |
| • Appropriate technology tools are available in meeting spaces. |
| • Workspace supports shifts between individual and collaborative work. |

included five variables, and explained 59 percent of the variation in the collaboration index. The predictor variables are listed in Table 5.9.

| Table 5.9 Variables Predicting Collaboration for Agents |
|---|
| • Space supports confidential conversations. |
| • Group members are supportive toward each other. |
| • Ease of accessibility to others. |
| • Ability to choose the physical location to do work. |
| • Office design conveys the appropriate corporate image. |

Table 5.9 shows that the predictor variables for collaboration for Agents include: space supports ability to have confidential conversations, group member supportiveness, accessibility to others, control over location of work, design of space conveys correct image to employees.

***Prediction of Job Control.*** Table 5.10 lists the question items that comprise the job control index.

For Agents, we found a regression equation that included six variables,

| Table 5.10 Job Control Index |
| --- |
| • Ability to choose the physical location to do work |
| • Amount of control over rate (speed) of work. |
| • Amount of control over making decisions |

and explained 53 percent of the variation in the job control index. The predictor variables are listed in Table 5.11 The predictors include: conversational privacy, noise, storage, knowledge of how to use technology, degree to which design enables sense of group membership, and retention.

| Table 5.11 Variables Predicting Job Control for Agents |
| --- |
| • Satisfaction with conversational privacy. |
| • Satisfaction with noise in overall work area. |
| • Satisfaction with storage. |
| • Workgroup possesses adequate knowledge of how to use technology tools. |
| • The design of workspace supports feeling of belonging to work group. |
| • The design of the office space supports retention. |

***Prediction of Community.*** Table 5.12 lists the individual question items that comprise the "workspace design communicates corporate culture" index.

| Table 5.12 Workspace Design Creates Community |
| --- |
| • The design of workspace contributes to sense of belonging. |
| • The design of workspace helps support feeling of belonging to work group. |
| • The design of the office space conveys the appropriate image. |
| • The design of the office space helps retain talented people. |

For Agents, the regression equation that predicts the degree to which design communicates corporate culture, included six variables, and explained 72 percent of the variation in this index. The predictor variables are listed in Table 5.13. These predictor variables include: arrangement of furnishings, noise levels, access to technology, group participation, interaction and control over location to do work.

| Table 5.13 Variables Predicting: <br> Workspace Design Creates Community |
| :--- |
| • Arrangement of work surface, storage, chair, computer, etc. in the workstation. |
| • Satisfaction with the overall noise. |
| • Ease of access to technology tools. |
| • Participation in decision-making. |
| • Amount of face-to face interaction. |
| • Control over physical location of work. |

## 5. Analysis of Call Center Agent Performance Metrics

***Definitions.*** Most organizations with Call Centers collect basic measures of Call Center Agent performance in terms of time and number of transactions per agent. This group of indices is generally referred to as Agent Performance Metrics. As part of this study we gathered a subset of these Agent Performance Scores indices in an effort to assess the relationships between them and other data that we collected.

We selected the following metrics, listed in Table 5.14, to use in our analyses. Included with each Agent Performance Scores definition is the manner in which we operationalized the calculation of each Agent Performance Scores measure.

To calculate our Agent Performance metrics, we gathered a monthly aggregate of the three Call Center Agent indices starting 3 months before the study began and through the entire 1 year course of the study, for a total of 16 months. We sorted the data by whether individuals were in the Experimental or Control group and computed an overall average monthly score for each group, by Call Center Agent, for each of the measures. We then determined which month each group had moved, in order to create a pre- and post-move monthly average for each group for each measure.

| | Table 5.14 Definitions of Agent Performance Metrics | |
| --- | --- | --- |
| | **Definition** | **Operationalized** |
| **ACD** | Automatic Call Distributor. The number of incoming calls received by each agent. | Measured as the monthly average of the number of incoming calls received per Call Center Agent, the monthly average per operator being the unit of analysis, over the course of the study. |
| **ACW** | After-Call Work. Work that is necessitated by and immediately follows an inbound transaction. The agent is unavailable to receive another inbound call while in this mode. | Measured as the monthly average of the seconds of time per transaction per Call Center Agent, the monthly average time in seconds per operator being the unit of analysis, over the course of the study. |
| **AHT** | Average Handle Time. The sum of Average Talk Time and Average After-Call Work for a specified period of time. | Measured as the monthly average of the seconds of time per transaction per Call Center Agent, the monthly average time in seconds per operator being the unit of analysis, over the course of the study. |

These averages, then, are the average per month for the months before the month of the move (pre-move) and per month for the months after the month of the move (post-move). Pre- to post-move average percent change scores were then computed for the Experimental and Control groups for each of the three Agent Performance Scores variables. The change scores were computed as actual change in numbers and percent change. The Agent Performance Scores results can be found in Table 5.15.

*Experimental Versus Control Group Differences.* There was no significant difference between the Experimental and Control group in their changes on ACD from pre- to post-move (see Table 5.15). Both groups handled fewer calls (per month, on the average) in the post-move time frame than in the pre-move timeframe, and the decline in calls handled was roughly the same proportion for the Experimental and Control groups.

In addition, there was no significant difference in the change in ACW (after-call work) time between the Experimental and Control groups from pre- to post-move (see Table 5.15).

There was a significant difference between the Experimental and Control groups in their change from before to after the move on AHT (average handle time) (see bottom row, Table 5.15). While AHT increased for both groups after the move, the Experimental group had a significantly greater change in AHT than did the Control group.

| Table 5.15 Change in Agent Performance Scores for Experimental and Control Groups -- Pre- and Post- Move | | |
|---|---|---|
| | **Experimental Group** | **Control Group** |
| | Post-Move Change from Baseline: # /% | Post-Move Change from Baseline: # # /% |
| ACD | -91 / -15% | -99 / -13% |
| ACW | 55 / +44% | 31 / +34% |
| AHT | 66 / +14% | 46 / +9% |

We examined the correlations between all participants' responses to individual survey items, and the pre to post-move percent change scores on the Agent Performance Scores measures. We found statistically significant correlations ($p < .01$) between changes in responses to some survey question items, and percent changes for the Agent Performance Scores measures of ACD (Automatic Call Distributor) and ACW (After-Call Work). Correlations show the strength of relationships between variables; they cannot be used to predict relationships or show causation between two variables.

Table 5.16 shows the percent change in ACD was related to changes in six survey question items. The first three question items are related to workstation design, and the following three items are related to work process and group identity issues. Table 5.16 shows that for the first three workstation-related question items, participants having increased satisfaction with workstation arrangement, size of work surface, or workstation storage from pre- to post-move, also tended to have increases in their average number of calls taken from pre- to post-move. This finding suggests that as individuals become more satisfied with these aspects of their space, they may be more efficient on their calls. It may be that their increased comfort with their workstation (layout, work surface size, or storage) means that they are better able to access and work with materials in a more seamless fashion during or following calls and therefore take more calls.

The data shows a negative relationship between increase in control over the pace of work, and number of calls taken (see Table 5.16). This suggests that if individuals are able to take calls at their own pace (presumably a pace at which they feel they are able to do a high quality job), they may end up taking fewer calls.

| Table 5.16 Relationship between Survey Measures and ACD Agent Performance Scores | |
|---|---|
| Survey Question Item | Correlation to Change in ACD |
| • Arrangement of work surface, storage, chair, computer, etc. in work-station. | .17 |
| • The size of the work surfaces. | .10 |
| • The amount of storage in workstation. | .11 |
| • Control over the rate (speed) of work. | -.09 |
| • Ease of shifting from individual to collaborative work. | -.09 |
| • The design of workspace supports feeling of belonging to work group. | -.13 |

Finally, the last two question items in Table 5.16 illustrates a negative relationship between the ability to quickly shift from individual to collaborative work, and feeling a part of the work group, and number of calls handled. Perhaps as Call Center Agents find it easier to help one another with their work and thus identify with their team, they are actually increasing the quality of their work. The other side of this is that helping one another, while it is likely to improve quality, takes time and may result in fewer calls taken.

Table 5.17 shows the percent change in ACW was correlated to changes in three survey items.

| Table 5.17 Relationship Between Survey Question Items and ACW Scores | |
|---|---|
| Question Item | Correlation to change in ACW |
| • Privacy provided by workstation. | .14 |
| • Satisfied with the noise. | .11 |
| • Workstation supports confidential conversations. | .15 |

All three of the items that were correlated with ACW are issues of noise and conversational privacy at the workstation. On two question items dealing with participants' satisfaction with conversational privacy at their worksta-tion, increases in their satisfaction from pre- to post-move were related to increases in ACW time. Increases in participants' satisfaction with the sound level in their work area were also related to increases in ACW time.

There are a couple of possible interpretations of these findings. One may be that individuals who feel that their personal work area supports their needs for quiet and privacy are able to do higher quality work. However, doing higher quality work may take more time. Another possible interpretation is that higher levels of privacy get in the way of working easily with others and working with others (e.g., getting help with questions, overhearing how others handle different situations, etc.) is important to efficient completion of ACW.

Overall, it appears that increasing some elements of workstation quality (layout, work surface size, and storage) may result in more calls taken, enhancing group work may result in decreasing the number of calls taken, and increasing workstation privacy may increase after-call work time. It is difficult to make firm conclusions or recommendations based on these findings because of the limitations of correlational data analyses, but these relationships are interesting because they do illustrate directional relationships between workstation features and capabilities, and changes in Agent Performance Scores.

## 6. Voluntary Separations

Data on voluntary separations from The Company were collected and summarized on a monthly basis throughout the study period. Separations were defined as those who left the company of their own accord (through retirements, other job opportunities, or other). It did not include involuntary terminations or job movements to other business units during the study. We divided the separated employees into groups based upon whether they belonged to the Experimental or Control group. These data are presented in Table 5.18.

| Table 5.18 Separations by Treatment Group | | |
|---|---|---|
| Separation Status | Experimental Group | Control Group |
| Not Separated | 94.7% | 91.9% |
| Separated | 5.3% | 8.1% |

Table 5.18 shows that over the course of the study, the Experimental group experienced 2.8 percent fewer separations than did the Control group. This suggests that there is a relationship between working in the new, more open work environment and reduced separations. We found that this rate of voluntary separations was consistent across departments as well.

## B. Conclusion

The primary purpose of this study was to determine the effects of the move to the new work environment on employee behavior and performance. The results of this study are completely congruent with research we have conducted with many other companies in which employees moved to more open work environments -- the tradeoff between enhanced collaboration (which has been stated as a key business objective of The Company) and reduced satisfaction with conversational privacy and noise.

This study has taken the additional step of linking the workplace with business outcomes, by collecting metrics on voluntary separations and Agent Performance Metrics, and incorporating this information into the analysis. We found that the move to the new environment had no effect on terminations (at least as defined in this study), but did have an effect on one Agent Performance Metric, AHT (Average Handle Time). We also found relationships between specific workplace design features and Agent Performance Metrics.

## VI. CASE STUDY: EFFECTS OF A BUILDING CONSOLIDATION ON EMPLOYEE BEHAVIOR AND BUSINESS PROCESS TIME AND COST

### A. Introduction

This Case Study discusses the results of the workplace evaluation that was conducted as part of the Workplace Consolidation project for the Headquarters campus of a leading company in the overnight shipping industry. The participants in this research are professional employees that support internal processes such as HR, Finance, Legal, and other activities typically found in corporate headquarters. The evaluation project had three major objectives:

1. Measure changes in employee behavior and reactions to the new work environment before and after the consolidation, by assessing:

- Fit of Design Features to work needs
- Within the workstation (lighting, storage, etc.)
- Overall layout of the space (support for training, etc.)
- Individual Work Process (interruptions, privacy, etc.)
- Group Work Process (collaboration, responsiveness, etc.)
- Employee Health
- Employee Satisfaction scores

2. Create "work environment -- employee -- work effectiveness models" that show a statistical links between workplace features (workstation design, the overall layout of the space, and the consolidation itself) and changes in employee behavior, on various measures of work effectiveness and work process improvements. We will also refer to these as "path models" in the results.

3. Create a financial benefits model that illustrates the financial impact of improvements to work process and other measures that resulted from the consolidation.

### B. Background of Consolidation/Re-Design Project

This project involved the consolidation of four office buildings (see Figure 5.14). The goals of the consolidation project were to: increase productivity due to a reduction in physical travel between buildings as well as decreasing travel time between departments, increase collaboration, increase communication, decrease approval process time and cost, and increase retention.

**Figure 5.14** View of two buildings on the campus (Author)

Prior to the consolidation, the employees within the 15 departments of the corporate headquarters were housed in a mix of office systems furniture "cubicles" with high panel walls and private offices, with employees and departments scattered throughout four buildings. (See Figures 5.15, 5.16, 5.17, and 5.18).

The consolidation and move was made to a new 137,000 square foot facility. To enhance employee communication within and across groups, the changes included moving to a more open environment in which very few employees retained a private, enclosed office (see Figure 5.18). Instead of private offices, or workstations with high panel walls, most employees were given workstations with 54" high wall partitions, and all employees were moved to a single building.

An important goal of the consolidation and redesign was to support efficient communication within departments and work groups, and between departments. To better support these types of collaboration, workstations were re-designed to better support working with others, and additional group meeting spaces were provided. Private offices were moved to the interior of the space to allow for natural daylight throughout the office space.

During the same time frame as the physical space redesign and consolidation, there were no other major changes to the IT systems used by the participating groups, nor were there any organizational changes, or changes to business processes. Thus, our analyses were able to focus on changes that we could attribute to the work environment (as opposed to technology or improved processes).

**Figure 5.15** Open plan workstations in the interior in clusters of six and eight (Author)

**Figure 5.16** View to private office (Author)

**Figure 5.17** Open plan offices on the perimeter window wall (Author)

**Figure 5.18** Open plan workstations in the new building (Author)

An important goal of the consolidation and redesign was to increase communication within and between departments and work groups by reducing the distance between people that frequently work together. In addition to reducing distances between people, the space was redesigned to optimize adjacencies between groups based on frequency of interaction, to provide workstations that support collaboration, and provide additional group meeting spaces.

## C. Research Methods

We employed three methods of assessment, including; a questionnaire given to all employees at three points in time, managers' ratings of the quality of output of each department (prior to and after the move), and a business process analysis (BPA) that included collecting data on time and costs for one business process that was identified as critical by the management team. Data was collected using this BPA method before and after the move-in to the new space.

### 1. Workplace Questionnaire

The questionnaire was administered to all headquarters employees three times: first, prior to the move to the redesigned space, a second time after employees had been in their new space for about 60 days, and finally a third time, 60 days after the second measurement. The purpose of administering the survey the third time was to test for the stability of any effects of changes in behavior we might observe.

The questionnaire included these sections:

- Background information on the respondent
- Office Design Issues
- Evaluation of Amenities
- Travel Time and Distances
- Privacy Issues
- Adjustability of Workstation Features
- Communication between Departments
- Communication between Individuals
- Job Satisfaction
- Physical Comfort and Health
- Stress Reactions
- Individual Performance
- Quality and Effectiveness of Work Group Process

Many of the individual question items used in the survey had already been statistically tested for reliability and used in other research projects. Others were created, tested, and employed specifically for this project. The topics listed above were measured by grouping together larger numbers of individual survey questions that "tap in" to each topic. A number of these "composite variables" were developed and used in the analysis of the findings. A composite variable is one in which scores on a number of different questions relate to the same subject. It is combined and averaged to form one stronger variable, or index, which gives an overall picture of how respondents felt about that topic.

## 2. Departmental Objectives Metrics

The purpose of these interviews was to identify the departmental business objectives, and quantifiable measures of performance of these objectives, from the leader of each department. Our purpose for these interviews was ultimately to relate changes in the workplace to improvements on the measures of these objectives. In the manager's interview, all 15 department managers were asked to describe the goals and objectives for their department. Follow up interviews were then conducted to create metrics (aspects of each goal that could be quantified in some way) and then data that represented current performance on each objective. These interviews were conducted before and after the consolidation and re-design. The data from these interviews was used in the financial model and in other parts of this project.

## 3. Business Process Analysis

A business process analysis (BPA) was conducted on the "Business Plan Approvals" process. This process involves the efforts and input of 8 of the 15 groups participating in this study, and is a complex process that was identified as being critical to the success of the organization. As part of the BPA, we broke the process into discrete activity/steps (see Figure 5.19). At each step we identified the resources (people or groups) required, time (in minutes) to complete the step, and tools used to complete the work. In addition, we identified other features of the process, such as re-work loops, wait times, and parallel sub-processes. We then worked with Human Resources to collect hourly "fully burdened" compensation rates for each resource. This data was used to calculate process cost at each step and a cost for the process overall.

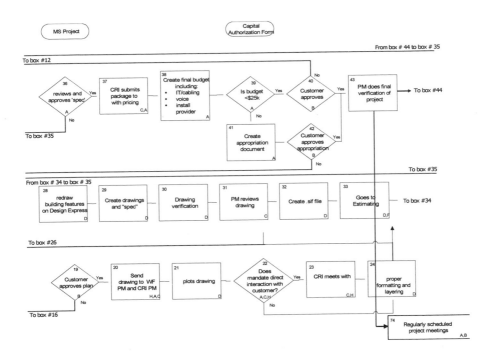

**Figure 5.19** Example portion of "Process Map" resulting from BPA (Author)

## 4. Design of the Study

The participants in this study were grouped into one of three treatment conditions: a Control group of employees who received neither the consolidation nor new furniture (this was all participants, prior to any change), employees who experienced the consolidation and redesign but no new furniture, and employees who received both the consolidation and redesign, and received new furniture (see Figure 5.20).

Figure 5.20 graphically illustrates these three conditions. One condition that was not tested as part of this study was employees who received new furniture "only," but were not a part of the consolidation and redesign (see empty cell, Figure 5.20).

**Consolidation and Re-design**
YES                              NO

|  | Consolidation<br>Only | No Consolidation<br>No new Furniture |
|---|---|---|
| **Received New Furniture** NO | | |
| **Received New Furniture** YES | Consolidation<br>And<br>New Furniture | |

**Figure 5.20** Design of study (Author)

The data was collected in three phases: once before any change to the environment and before any move, and twice after the move. Figure 5.21 shows the data collection points ("O" stands for "observation") and the timing of the move and consolidation (signified by the "X").

$$O_1 \quad X \quad O_2 \quad O_3$$

**Figure 5.21** Data collection plan (Author)

**D. Sample Characteristics**

1. Response Rates

Table 5.19 shows that a total of 1275 surveys were distributed and 721 completed surveys were returned, for a respectable 57 percent overall completion rate (17 to 30 percent is a typical return rate for survey studies).

| Table 5.19 Response Rates for Each of the 3 Data Collection Points | | | | |
|---|---|---|---|---|
| | Pre-change | Post-Change 1 | Post-Change 2 | Total |
| Surveys Distributed | 425 | 425 | 425 | 1275 |
| Surveys Returned | 269 | 257 | 195 | 721 |

2. Participant Characteristics

In this section we describe the general characteristics of the population we sampled, including: gender, job description and departmental affiliation. These statistics are stated by percentage of the overall sample of 721 respondents. Table 5.20 shows a breakdown of all participants by gender. Participants in this sample are predominantly female (62 percent).

| Table 5.20 Participants by Gender | |
|---|---|
| **Gender** | **Percent Participating** |
| Female | 62 |
| Male | 38 |

Table 5.21 shows the breakdown of participants in this study by job category. The greatest proportion of participation is by Professionals (45 percent) and General Office (23 percent) which is representative of the proportions of the actual population.

Table 5.22 shows a breakdown of participants by department. This table shows that there is no one department that dominates the analysis, and that proportionally, participation is representative of the actual population.

**E. Results**

1. Variables Analyzed

In this section the analyses are divided into the following six categories:

- Fit of Workspace Design Features to work needs
- Overall Layout of the space (support for training, etc.)
- Individual Work Process (interruptions, privacy, etc.)
- Group Work Process (collaboration, responsiveness, etc.)
- Employee Health

- Job Satisfaction scores

| Table 5.21 Participants by Job Type | |
|---|---|
| Job Description | Percentage of Total |
| Professional (accountant, computer programmer, engineer, marketing specialist) | 45 |
| Technical (customer invoicing, collections, credit, etc.) | 16 |
| General Office (coordinator, assistant, secretary) | 23 |
| Manager/Senior Manager | 10 |
| Managing Director | 4 |
| VP/Senior VP | 2 |

| Table 5.22 Participants by Department | |
|---|---|
| Department Name | Percentage of Total |
| Marketing | 9 |
| Customer Service | 4 |
| Legal | 2 |
| Finance | 4 |
| IT | 10 |
| HR | 13 |
| Sales | 5 |
| Accounting/Purchasing | 15 |
| Collections | 8 |
| Invoicing | 10 |
| Ground Operations | 1 |
| Service Systems/Planning and Operations | 10 |
| Security | 1 |
| Other | 8 |

## 2. Method of Analysis

Participant responses to each variable were compared on the three study conditions ("before change," "consolidation only," and "consolidation and new furniture"). The overall statistical significance of each of these compar-isons was computed using an analysis of variance (ANOVA) program, and a Bonferroni's *post hoc* test to detect differences between the three conditions was calculated. In all cases the minimum criteria for statistical significance being set at $p < .05$. In all cases reported in this section, the differences are statistically significant. This means we can be sure, at a 95 percent level of confidence, that the percentage of change shown on each evaluation repre-sents an actual change in participant's attitudes.

Table 5.23, and all other Tables showing results information in this section, use the same format for displaying the results. The column headings at the top of the Table show the conditions "Prior to Changes" and "Consolidation Only." The name of each variable is shown on the leftmost column of the Table. Immediately to the right of each variable name, the conditions of "Consolidation Only" and "Consolidation and New Furniture" are shown. A matrix is formed between the row and column conditions, showing the three possible comparisons.

For instance, in Table 5.23, for the "Lighting" feature, the results show a 12 percent increase in employees' perceptions of the suitability of lighting in their workstation when we compared their perceptions "Prior to Changes" with those of employees who received the "Consolidation Only." When comparing employee perceptions prior to changes, with those receiving the consolidation and new furniture, we find a 15 percent increase in perceptions of suitability of lighting (see Table 5.23). We found no change in perceptions between employees who received only the consolidation with those who received consolidation and new furniture (see Table 5.23). This format allows us to quickly look for patterns in responses to the different conditions.

## 3. Effects by Job Type

We conducted analysis for each of the variables to determine if job type plays a role in employee evaluations of the work environment, individual work process (perceptions of interruptions, etc.), group work process (col-laboration, etc.), health, and Job Satisfaction scores. The results showed virtually no main effects for job type, meaning that employees across job types had very consistent responses to the majority of issues we investigated. Thus for the analyses we present in this report, we report responses for all employees as a group and do not present analyses by job type. This has the positive side effect of greatly simplifying both the presentation and interpretation of the results and their implications to the organization.

4. Fit of Workstation Design Features to Work Needs

In this section we describe the results of our analyses of the fit of the work environment to employee work needs. In this analysis, we divide the work environment by scale of size: 1) interior features of the workstation, and 2) larger scale design features of the floor plan and interior layout and features of the building.

Table 5.23 shows a summary of the impacts of the consolidation and redesign, and the consolidation with new furniture on employees' perceptions of workstation features and overall layout of the new space. For both evaluations, the comparisons for the consolidation, and consolidation and furniture, are compared against the baseline of employee assessments before any change. We also compared the effects of the consolidation/redesign against the consolidation/redesign with new furniture. In Table 5.23, we summarize the findings for workstation design features, including: Lighting, Glare Reduction, Air Quality, Frequency of Adjustment, Storage, Workstation Layout, and Comfort.

Overall, the investment in the consolidation and redesign of space has brought consistent, significant improvements in employee evaluations of the fit of the workspace to their work process and satisfaction. The addition of new furniture indicates a trend of improvement, particularly with increased frequency of adjustment, in which employees more frequently adjust or move features within their workstations, to support their work process.

*Suitability of Lighting.* Employees who received the consolidation/redesign, and the consolidation coupled with new furniture, reported improvements in lighting ranging from 12 to 15 percent (respectively) when compared to evaluations before any changes (see Table 5.23). However, the results show that there was no difference in suitability of lighting between employees who received the consolidation, and those who received the consolidation and new furniture.

*Reduction in Glare and Shadows.* Employees who received the consolidation/redesign only, reported a 15 percent reduction in glare and shadows when compared to employee assessments of glare before any changes (see Table 5.23). There was no reduction in glare and shadows between employees who received the consolidation and furniture when compared to assessments of glare before any changes (see Table 5.23). There was no reduction in glare and shadows between employees who received the consolidation only, and those who received the consolidation with new furniture. Redesign of space, rather than new furniture, has the biggest impact on glare reduction.

| Table 5.23 Impact of Treatment Conditions on Workspace Evaluations | | | |
|---|---|---|---|
| All percentage results listed are statistically significant. "NS" indicates that there is no statistically significant difference between the groups being compared. | | | |
| | **Treatment Condition** | | |
| **Issue Evaluated** | Consolidation and Redesign Only condition, compared to "Prior to Changes" | Consolidation with New Furniture compared to "Prior to Changes" | Consolidation with New Furniture compared to "Consolidation and Redesign Only" |
| Lighting | 12% | 15% | NS |
| Glare Reduction | 15% | NS | NS |
| Air Quality | 28% | 26% | NS |
| Frequency of Adjustment | NS | 49% | 34% |
| Storage | 14% | NS | -13% |
| Workstation Layout | 8% | 14% | NS |
| Comfort | 13% | 18% | NS |

*Air Quality.* Employees who moved to the consolidated spaces, and those who moved to the consolidated spaces coupled with new furniture, reported improvements in air quality ranging from 28 to 26 percent (respectively) when compared to evaluations before any changes (see Table 5.23). However, the results show that there was no significant difference in air quality between employees who received the consolidation only, and those who received the consolidation and new furniture. Thus, as one would expect, the addition of new furniture itself did nothing to affect air quality. The redesign, resulting in more open space, had the biggest impact on air quality.

*Frequency of Adjustment of Workstation Features.* Employees who received the consolidation only reported no increase in adjustment of workstations (see Table 5.23). However, the results show that there was a significant increase in adjustment for employees who received the consolidation and new furniture, over employees who had no changes, and over employees who received just the consolidation (see Table 5.23). Clearly the redesign and redeployment of furniture, and use of new furniture, were consistently responsible for increased adjustability -- and logically the consolidation had no effect on adjustability.

*Storage*. Employees who received the consolidation/redesigned space reported a 14 percent increase in their ratings of the quality of storage when compared to employee assessments of storage before any changes (see Table 5.23). There was no improvement in storage among employees who received the consolidation/redesign with new furniture when compared to assessments of storage before any changes (see Table 5.23). There was a significant 13 percent reduction in assessed quality of storage between employees who received the consolidation with new furniture, and those who received only the consolidation. Thus, it appears that the group that received new furniture had some reduction in storage capacity (particularly the overhead storage bins) that employees who got only redesigned space did not. This finding serves to reinforce the point that good user analysis and design are fundamental to creating useful space, and new furniture cannot overcome the effects of design decisions that do not support user needs.

*Workstation Layout*. Workstation layout was defined through a number of measures, including: work materials close at hand, floor area, arrangement of equipment and furnishings, work surface size, ease of organizing work materials. Employees who received the consolidation/redesign only, and the consolidation coupled with new furniture, reported improvements in workstation layout ranging from 8 to 14 percent (respectively) when compared to evaluations before any changes (see Table 5.23). The results also show that there was no difference in satisfaction with workstation layout between employees who received the consolidation, and those who received the consolidation and new furniture. Thus, satisfaction with workstation layout is dependent upon quality of design, and not through simply giving workers new furniture.

*Comfort*. Employees who received the consolidation/redesign only, and the consolidation coupled with new furniture, reported improvements in workstation comfort ranging from 13 to 18 percent (respectively) when compared to evaluations before any changes (see Table 5.23). The results show that there was no difference in comfort between employees who received the consolidation/redesign and those also receiving new furniture. This finding points out the importance of good design and interior layout of space, along with seating appropriate to the task, to ensure employee comfort. New furniture must be leveraged with design to produce positive results.

Overall the investment in the consolidation/redesign, and subsequent improvements in adjacencies, have brought consistent, significant improvements in employee evaluations of the fit of the workspace to their work process and satisfaction.

5. Impact of Large Scale Design Features on Meetings and Travel Time

In this section, we analyzed employees' perceptions of the fit between the large-scale design features associated with the consolidation and layout of the building, and their work process needs. Tables 5.24 and 5.25 summarize our analysis for large-scale interior design and layout issues, including: training spaces, travel time, and support for group meetings.

*Support for Meetings Between Work Groups.* Employees who received the consolidation/redesign only, and the consolidation coupled with new furniture, reported significantly greater support for meetings between work groups, ranging from 23 to 26 percent (respectively) when compared to evaluations before any changes (see Table 5.24). The results show that there was no difference on this issue between employees who received the consolidation and new furniture, and those who received only the consolidation. This is not surprising because this is a floor plan layout and adjacencies issue, not a workstation issue.

*Support for Meetings Between Members of the Same Work Group.* Employees who received the consolidation/redesign only, and the consolidation coupled with new furniture, reported significantly greater support for meetings between work groups, in both cases 35 percent when compared to evaluations before any changes (see Table 5.24). The results show that there was no difference on this issue between employees who received the consolidation/new furniture, and those who received only the consolidation. This is not surprising because this is a floor plan layout and adjacencies issue, not a workstation issue.

*Availability of Spaces for Training.* Employees who received the consolidation only, and the consolidation coupled with new furniture, reported significantly greater availability of spaces for training, ranging from 30 to 36 percent when compared to evaluations before any changes (see Table 5.24). The results show that there was no difference on this issue between employees who received the consolidation/new furniture, and those who received only the consolidation.

*Satisfaction with Distance to Amenities.* Employees who received the consolidation/redesign only, and the consolidation coupled with new furniture, reported significantly reduced satisfaction with their distance to amenities, ranging from 33 to 28 percent less when compared to their satisfaction before any changes (see Table 5.24). The results show that there was no difference on this issue between employees who received the consolidation/new furniture, and those who received only the consolidation.

*Travel Time between Airport and Workstation.* Employees who received the consolidation only, and the consolidation coupled with new furniture, reported significantly reduced travel time between airport and workstation, ranging from 23 to 33 percent less when compared to their travel times before any changes (see Table 5.24). The results show that there was no difference on this issue between employees who received the consolidation/new furniture, and those who received only the consolidation.

*Travel Time (Waiting for the Elevator).* Employees who received the consolidation only, and the consolidation coupled with new furniture, reported significantly reduced time spent waiting for the elevator, ranging from 59 to 45 percent less when compared to waiting times before any changes (see Table 5.24). The results show that there was no difference on this issue between employees who received the consolidation/new furniture, and those who received only the consolidation.

*Travel Time (Total Time from Home to Work).* Employees who received the consolidation/redesign only, and the consolidation coupled with new furniture, reported significantly increased travel time from home to work, ranging from 29 to 32 percent more when compared to travel time before any changes (see Table 5.24). The results show that there was no difference on this issue between employees who received the consolidation/new furniture, and those who received only the consolidation.

*Travel Time (Time to Travel from Workstation to All Other Departments).* Employees who received the consolidation/redesign only, and the consolidation coupled with new furniture, reported significantly reduced travel time from workstation to other departments, in both cases 65 percent less travel time than before any changes (see Table 5.24). The results also show that there was no difference in travel time between employees who received the consolidation and new furniture, and those who received only the consolidation. Thus travel time was dependent on the departmental adjacencies, not the use of new furniture.

## 6. Individual Work Process

In this section we describe the results of our analyses of the impact of consolidation and new furnishings on various aspects of individual work process, including: ability to have confidential conversations, distractions, interruptions, handling confidential material, and privacy. Table 5.25 summarizes these findings.

*Confidential Conversations in Workstation.* Employees who received the consolidation/redesign only, and the consolidation coupled with new furniture, reported no difference in their ability to have confidential conversations when compared to evaluations before any changes (see Table 5.25). The results show that there was no difference in ability to have confidential conversations between employees who received the consolidation and new furniture, and those who received only the consolidation. Thus despite the overall trend toward greater openness in the workstations after the consoli-

| Table 5.24 Summary of Support for Meetings and Travel | | | |
|---|---|---|---|
| All percentage results listed are statistically significant. "NS" indicates that there is no statistically significant difference between the groups being compared. | | | |
| | **Treatment Group** | | |
| | Consolidation and Redesign Only condition, compared to "Prior to Changes" | Consolidation with New Furniture compared to "Prior to Changes" | Consolidation with New Furniture compared to "Consolidation and Redesign Only" |
| Support for Meetings between Work Groups | 23% | 26% | NS |
| Meetings between Team Members | 35% | 35% | NS |
| Spaces for Training | 30% | 36% | NS |
| Satisfaction w/ Distance to Amenities *(fitness center, restaurant, parking, shopping, banking machine)* | -33% | -28% | NS |
| Travel time: Airport to workstation | -23% | -33% | NS |
| Travel Time: Waiting for elevator | -59% | -45% | NS |
| Travel Time: Home to Work | 29% | 32% | NS |
| Travel Time: Workstation to all Departments | -65% | -65% | NS |

dation and with the new furniture there was no reduction in employees' ability to have confidential conversations in their workstations (see Table 5.25).

***Visual and Noise Distractions in Workstation.*** Employees who received the consolidation/redesign only, and the consolidation coupled with new furniture, reported no difference in problems with visual and noise distractions in their workstations when compared to evaluations before any changes to the work environment (see Table 5.25). The results show that there was no difference in distractions between employees who received the consolidation and new furniture, and those who received only the consolidation. Thus, despite the overall trend toward greater openness in the workspaces after the consolidation and with the new furniture, there was no increase in problems with distractions within the workstation (see Table 5.25).

*Interruptions in the Workstation.* Employees who received the consolidation/redesign only, reported a 13 percent reduction in interruptions in their workstations when compared to employee evaluations before any changes to the work environment (see Table 5.25). There was no difference in level of interruptions between employees who received the consolidation and new furniture when compared to evaluations before any change to the work environment (see Table 5.25). There was no difference between employees who received the new furniture and those received only the consolidation. This shows the importance of proper programming and design in reducing interruptions. Furniture by itself cannot prevent interruptions -- it must go hand in glove with design.

*Handle Confidential Material in Workstation.* Employees who received the consolidation only, and the consolidation coupled with new furniture, reported no difference in ability to handle confidential material in their workstation when compared to evaluations before any changes to the work environment (see Table 5.25). The results show that there was no difference in ability to handle confidential materials between employees who received the consolidation and new furniture, and those who received only the consolidation.

Thus, despite the overall trend toward greater openness in the workspaces after the consolidation and with the new furniture there was no decrease in ability to handle confidential materials within the workstation (see Table 5.25).

| Table 5.25 Summary of Changes to Individual Work Process | | | |
|---|---|---|---|
| All percentage results listed are statistically significant. "NS" indicates that there is no statistically significant difference between the groups being compared. | | | |
| | Treatment Group | | |
| | Consolidation and Redesign Only condition, compared to "Prior to Changes" | Consolidation with New Furniture compared to "Prior to Changes" | Consolidation with New Furniture compared to "Consolidation and Redesign Only" |
| Confidential Conversations | NS | NS | NS |
| Visual and Noise Distractions | NS | NS | NS |
| Interruptions | -13% | NS | NS |
| Handle Confidential Material | NS | NS | NS |
| Privacy | +8% | NS | NS |

***Privacy in the Workstation.*** Employees who received the consolidation/ redesign reported an 8 percent increase in privacy in their workstations when compared to employee evaluations before any changes to the work environment (see Table 5.25). However, there was no change in level of privacy for employees who received the consolidation with new furniture when compared to evaluations before any change to the work environment, or when compared to the "consolidation only," group (see Table 5.25). Overall these findings suggest the consolidation played a positive role in increased privacy - and that new furniture (and the lower amount of enclosure that came with it) added no additional gains in privacy but caused no decrement in privacy, either.

## 7. Group Work Process

In this section we describe the results of our analyses of the impact of consolidation and new furnishings on various aspects of group work process, including: communication between departments, departmental collaboration, departmental responsiveness, face-to-face collaboration, quality of group process, and group effectiveness. Table 5.26 summarizes these findings.

*Frequency of Communication Between Departments.* Table 5.26 shows that there was no difference in frequency of communication between departments on any of the comparisons. Thus the consolidation and new furniture did nothing to increase the frequency of communication. This is not, however, a measure of the ease or quality of that communication, for which other analyses (see Table 5.26) did show an improvement.

*Departmental Collaboration.* Employees who received the consolidation/ redesign only, and the consolidation coupled with new furniture, reported significantly increased collaboration between departments, ranging from 8 to 11 percent improvements on "cooperation, flow of ideas, getting in touch with each other" before any changes to the work environment (see Table 5.26). The results also show that there was no difference in departmental collaboration between employees who received the consolidation and new furniture, and those who received only the consolidation. Thus, the level of departmental collaboration was dependent on the consolidation, and not the use of new furniture, a logical finding since most collaboration is supported in the context of the overall layout of the space, not within a workstation.

*Face-to-Face Collaboration.* Employees who received the consolidation/ redesign only, reported significantly increased face-to-face collaboration between departments, an 8 percent improvement before any changes to the work environment (see Table 5.26). The results also show that there was no difference in face-to-face collaboration between employees who received the consolidation and new furniture, and employees before any changes (see Table 5.26). There was no difference in face-to-face collaboration between those who received only the consolidation, and those who received consolidation and new furniture. Thus, face-to-face collaboration was most influenced by the consolidation and improved adjacencies.

*Quality of Group Process.* Table 5.26 shows that there was no difference in self-assessed quality of group process on any of the comparisons. Thus the consolidation and new furniture did not affect the quality of group process.

*Group Effectiveness.* Employees who received the consolidation/redesign and new furniture, reported a 5 percent improvement in the quality and accuracy of their reports when compared to employees before any changes to the work environment (see Table 5.26). Thus, the combined features of consolidation and new furniture together enhanced group effectiveness.

*Departmental Responsiveness.* Employees who received the consolidation/redesign only, and the consolidation coupled with new furniture, reported significantly increased departmental responsiveness (as defined by

| Table 5.26 Summary of Changes to Group Work Process | | | |
|---|---|---|---|
| All percentage results listed are statistically significant. "NS" indicates that there is no statistically significant difference between the groups being compared. | | | |
| | **Treatment Group** | | |
| | Consolidation and Redesign Only condition, compared to "Prior to Changes" | Consolidation with New Furniture compared to "Prior to Changes" | Consolidation with New Furniture compared to "Consolidation and Redesign Only" |
| Frequency of Communication between Departments | NS | NS | NS |
| Collaboration and Workflow between Departments (cooperation, flow of ideas, getting in touch, delays in work) | 8% | 11% | NS |
| Face-to-face Collaboration | 8% | NS | NS |
| Quality of Group Process | NS | NS | NS |
| Group Effectiveness (quality and accuracy of reports) | NS | 5% | 5% |
| Departmental Responsiveness (ease of getting questions answered, speed of approvals) | 32% | 31% | NS |

getting questions answered and speed of approvals), ranging from 31 to 32 percent improvements before any changes to the work environment (see Table 5.26). The results also show that there was no difference in departmental responsiveness collaboration between employees who received the consolidation and new furniture, and those who received only the consolidation (Table 5.26) shows only a 1 percent difference between the two groups). Thus, departmental responsiveness was largely dependent on the consolidation and improved adjacencies, and less on the use of new furniture.

8. Employee Health

In this section we describe the results of our analyses of the impact of consolidation/redesign and new furnishings on various aspects of employee health. These measures of health include: discomfort (back pain, pain in arms and hands, visual discomfort) signs of psychological stress, and self-assessed individual health. Table 5.27 summarizes these findings.

*Back Pain.* Employees who received only the consolidation/redesign reported a 9 percent decrease in back pain when compared to employees before any changes to the work environment (see Table 5.27). There were no differences on any other comparisons (see Table 5.27).

*Discomfort in Arms and Hands.* Employees who received only the consolidation/redesign reported a 10 percent decrease in discomfort in arms and hands when compared to employees before any changes to the work environment (see Table 5.27). There were no differences on any other comparisons (see Table 5.27).

*Visual Discomfort.* Employees who received only the consolidation reported a 6 percent decrease in visual discomfort when compared to employees before any changes to the work environment (see Table 5.27). There were no differences on any other comparisons (see Table 5.27).

*General Health.* Employees who received the consolidation/redesign with new furniture reported a 13 percent increase in assessment of their overall health when compared to employees who received only the consolidation (see Table 5.27). There were no differences on any of the other comparisons.

*Psychological Stress.* There were no differences in signs of psychological stress on any of the comparisons (see Table 5.27). Thus, regardless of the major changes in location, interior redesign of overall space and workstations, and new furnishings, we detected no increased problems in self-reported measures of psychological stress.


9. Employee Job Satisfaction Scores

In this section we describe the results of our analyses of the impact of consolidation and new furnishings on employee pride in working for the company. We used the following question: "I feel proud to work for The Company." We view this as a proxy question for employee satisfaction, thus employee responses to this issue have important implications for the organization. We found a significant increase in employee agreement with this question (24 percent) for employees who experienced the consolidation/redesign. It seems reasonable to believe that employees moving to a new

| Table 5.27 Summary of Impacts to Employee Health | | | |
|---|---|---|---|
| All percentage results listed are statistically significant. "NS" indicates that there is no statistically significant difference between the groups being compared. | | | |
| | **Treatment Group** | | |
| | Consolidation and Redesign Only condition, compared to "Prior to Changes" | Consolidation with New Furniture compared to "Prior to Changes" | Consolidation with New Furniture compared to "Consolidation and Redesign Only" |
| Back Pain | -9% | NS | NS |
| Arm and Hand Discomfort | -10% | NS | NS |
| Visual Discomfort | -6% | NS | NS |
| General Health | NS | NS | +13% |
| Psychological Stress | NS | NS | NS |

location (through the consolidation) would feel positive about the fact that the company is investing in their work environment, and work efficiency. This could influence an increase in pride in their company, in satisfaction, and contribute to employee retention.

10. Path Models

*Explanation of Path Models.* A "path model," is a statistical model showing the predictive relationships between a group of variables (such as design features) and an outcome variable (which can be a measure of a behavior or work process). We created four path models to show the design elements that predict variables that were of key interest to this study. These path models include: privacy, group communication, group effectiveness, and business case Approval Time. The purpose of creating these path models is to create strategic information about work process and design feature issues that can be generalized to all areas of the organization. Figure 5.22 shows the path model for the privacy variable.

To the left of the word "Privacy" are five variables that we found predict employee's perceptions of privacy. Next to the arrow connecting each variable to Privacy is a number. This number can range from 0 to 1, and indicates the relative strength of the contribution of each variable to influencing perceptions of privacy. For privacy, "ability to handle confidential materials" while in the workstation was the strongest predictor, and "storage" the weakest. These numbers are for understanding relative contributions only, as all variables included within a model are important. Above the word "privacy" is the number "72%." This 72 percent is the total amount of variance in a given privacy score that we can attribute to the combined effects of the five variables (see Figure 5.22).

*Privacy Model.* We found that five variables predicted privacy, including: ability to handle confidential materials in the workstation, level of interruptions, ability to have confidential conversations in the workstation, interior layout of workstation (work materials close at hand, floor area, arrangement of equipment and furnishings, work surface size, ease of organizing work materials), and storage (sufficient shelves and filing in workstation) (see Figure 5.22).

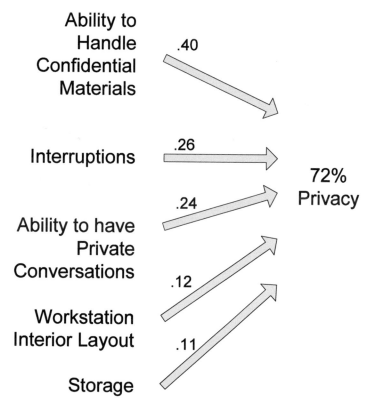

**Figure 5.22** Path Model for Privacy (Author)

Surprisingly, the level of enclosure (walls or panels surrounding the workstation) was *not* a predictor of privacy. This is an interesting model because all of the predictor variables are related to support of individual work process, rather than features like enclosure.

***Collaboration and Workflow between Departments.*** We defined "Collaboration and Workflow between Departments" through five measures, including: cooperation between departments, ease of flow of ideas, ease of getting in touch with members of other work groups, and delays in workflow between groups. We found that four variables predicted collaboration, including: overall lighting in the workstation (overhead, task, and daylight), travel time between the workstation and all other departments, amount of interruptions experienced within the workstation, interior layout of the workstation (work materials close at hand, floor area, arrangement of equipment and furnishings, work surface size, ease of organizing work materials), and storage (sufficient shelves and filing in workstation) (see Figure 5.23).

For the "Collaboration and Workflow between Departments" factor, all five predictor variables contributed about equally to influence Collaboration. The four predictor variables account for about 13 percent of the variance in departmental collaboration (see Figure 5.23).

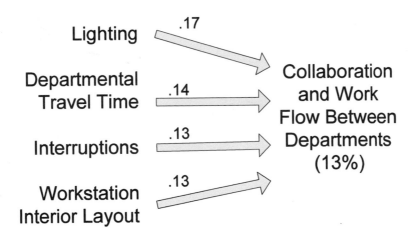

**Figure 5.23** Path Model for collaboration between Departments (Author)

While this is a small amount, it is still significant because it shows us the environmental features that can be manipulated to affect collaboration. Obviously there are many other variables such as job and organizational design, management practices, technology design, etc. that also influence collaboration. To enhance group collaboration and workflow between departments through workspace design, this analysis shows that resources should be allocated to optimize lighting and workstation layout needs, and design of proper adjacencies to reduce travel time between departments.

*Group Effectiveness.* We were interested in knowing if "Collaboration and Workflow between Departments" predicted any aspect of work performance. We created and tested a model that shows collaboration as we defined it in fact predicts 12 percent of the variance in Group Effectiveness (see Figure 5.24). Group effectiveness itself was measured as an index of work group performance on the following dimensions: problem solving, making decisions, getting the work done, making use of member skills, accuracy and quality of reports, developing new ideas, and overall effectiveness of the work group. Thus, if the design features that we discussed for the Collaboration model are enhanced, not only will collaboration and work flow be improved, but group effectiveness (as defined) will be positively impacted as well. While the 12 percent impact is a relatively small amount, it is still significant because it shows us the environmental features that can be manipulated to affect collaboration and ultimately effectiveness. To enhance group collaboration and workflow between departments by using the work environment, we recommend that resources be allocated to understanding lighting and layout needs, and design of proper adjacencies to travel time between departments is minimized.

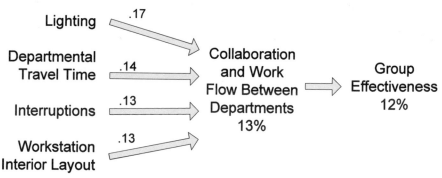

**Figure 5.24** Path model predicting Group Effectiveness (Author)

*Business Case Approval Time.* We measured "Business Case Approval Time" by conducting a business process analysis (BPA) of this process, and collecting data on the time and cost of conducting this process from key management leads in four departments. With this analysis, we assessed the amount of time (in days) it took to have a business case approved, before and after the consolidation. Thus, this is a measure of the efficiency of a business level process across departments. We found that five variables predicted a large and significant amount of the variance in speed of the approval process (53 percent), listed in order of importance: adequate spaces for meetings between groups, interior layout of the workstation (work materials close at hand, floor area, arrangement of equipment and furnishings, work surface size, ease of organizing work materials), level of visual and noise distractions, travel time between workstation and all other departments, and workstation has adequate space to support collaboration with another person (see Figure 5.25). Given this model, to enhance speed of the approval process, we recommend that resources be allocated to supporting meetings between groups, and within workstations. Careful adjacency planning should be used to minimize distance between departments that need to collaborate on various processes. Finally, interior workstation layout should be optimized.

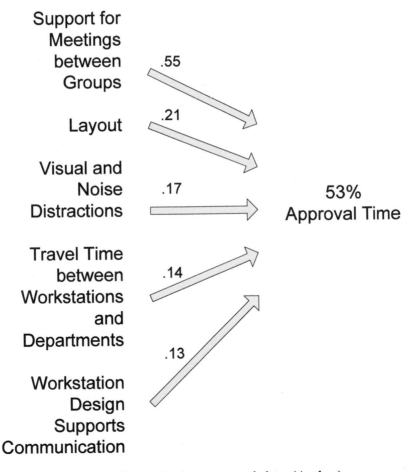

**Figure 5.25** Path Model predicting approval time (Author)

## 11. Business Process Analysis

A business process analysis (BPA) was conducted on the "Business Plan Approvals" process. This process involves the efforts and input of 8 of the 15 groups participating in this study, and is a complex process that was identified as being critical to the success of the organization. As part of the BPA, we broke the process into discrete activity/steps. At each step we identified the resources (people or groups) required, time (in minutes) to complete the step, and tools used to complete the work. In addition, we identified other features of the process, such as re-work loops, wait times, and parallel sub-processes. We then worked with Human Resources to collect hourly "fully burdened" compensation rates for each resource. This data was used to calculate process cost at each step and a cost for the process overall.

The time required for Approval Processes, which is a cross-departmental process involving departments including Legal, Finance, Properties, and IT, was reduced 32 percent after the consolidation. In addition, all departments' ratings of each other on "responsiveness," (getting questions answered, etc.) increased by 32 percent after the changes to the workspace.

This 32 percent time reduction translated into an annualized cost savings of $120,000 based on compensation data for roles and the departments involved in this process. There are numerous other cross departmental business processes occurring that were not measured, but the assumption is there are also time and (opportunity and dollar) cost savings far beyond what we measured. This research allowed us to pinpoint the exact issues that predict this time improvement and cost reduction in the Approval Process. These issues include: adequate space for meetings between groups, workstation interior layout that supports work process, reduced visual and noise distractions, reduced travel time between departments, and space within the workstation to support collaboration.

## D. Conclusions

Based on the research conducted with this organization, we found consistent positive impacts for the consolidation and redesign of spaces that reduced time and cost of travel between locations, ease of communication and collaboration, and even had significant effects on the time and cost of at least one ongoing business process. This study suggests that there is an enormous potential to increase the overall effectiveness and efficiency of collaborative behavior and related business processes in which decisions must be made across departments, anywhere in the company -- through use of optimized facility design. While the simple consolidation of buildings brings several immediate benefits, it is not the complete answer. While consolidation brings people together, the real power is with the proper design of adjacencies between functional groups, and design of space to support meetings within workstations and in the overall space. This study also provides insights as to the specific design features predictors that predict key perceptions, behaviors and efficiency outcomes. While much more work in this area needs to be done before the results can be generalized beyond this Case Study, these analyses show that workplace design is indeed related to performance and financial outcomes related to business process.

## VII. CASE STUDY: EFFECTS OF WORK ENVIRONMENT DESIGN ON EMPLOYEE BEHAVIOR AND BUSINESS PROCESS: THE MARKETPLACE BUILDING

This Case Study discusses the results of the workplace evaluation that was conducted as part of the workplace consolidation project for a manufacturer in the Midwest. The focus of this project is a new facility referred to as the "marketplace."

### A. Research Objectives

This evaluation project had three objectives:

1. To assess the effects of changing individual workspaces and overall architectural interior space on employee behavior, work practices, group process, departmental outputs, and selected business processes before and after the consolidation. To make strategic recommendations based on these results in terms of broader organizational implications.

2. To create regression models that show statistical linkages between workplace features (workstation design, the overall layout of the space, and the consolidation itself), changes in employee behavior, and various measures of work effectiveness and work process improvements.

3. To understand the financial impact of improved business process efficiency that resulted from the consolidation.

### B. Consolidation Project Objectives

The Marketplace project involved the consolidation of employees from facilities in five different locations. The goals of the consolidation project were to:

♦ create a new concept for work environments that enable higher worker effectiveness
♦ increase communication and collaboration between individuals and groups
♦ create operational efficiencies through business process effectiveness
♦ increase productivity due to a reduction in physical travel between buildings as well as decreasing travel time between departments
♦ increase job satisfaction, and sense of belonging
♦ communicate corporate identity and image to customers

A model was created for understanding the business drivers of the organization, which also assisted in guiding decisions concerning the layout, interior design and furnishings of the new workspace. This model is shown in Figure 5.26.

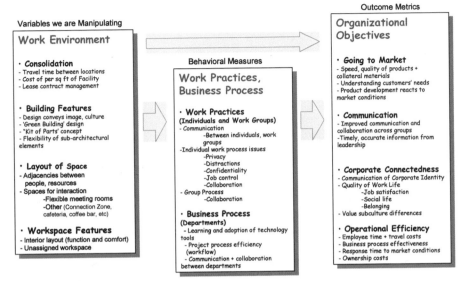

**Figure 5.26** Strategic Business Model (Author)

Prior to the consolidation, employees and departments were located in five different buildings, each having a different mix of traditional office systems furniture and space allocations for individual workspace, private offices, and collaborative meeting spaces.

As part of this project, a consolidation was conducted in which employees from five locations were moved into the Marketplace building. In the Marketplace, the furniture infrastructure and interior layout were designed to optimize collaboration and communication between and across groups. In the new space, few employees retained a private, enclosed office. Instead of private offices or workstations with high panel walls, all employees were given workstations with a high degree of openness (see Figure 5.27). In addition, there are no private offices, and employees were assigned either a dedicated or shared workspace based on the type of work style of each employee. Rather than the traditional 1:1 ratio of employees to workstations, the Marketplace project uses a 1 workstation to 1.5 people ratio. Employees who share a workstation are assigned docking stations to use for their personal items, storage, and as a place to receive information (see Figure 5.28). Compared to the buildings employees were previously located in, a greater amount of space was dedicated to collaborative areas and privacy rooms (see Figures 5.29 and 5.30).

**Figure 5.27** Open Workstations (Author)

Based on the open plan concept, colors were used to help establish boundaries and to assist in way-finding throughout the building (see Figure 5.32).

**Figure 5.28** Docking Stations (Author)

**Figure 5.29** Collaboration spaces (Author)

**Figure 5.30** Private Office (Author)

In addition, the architecture of the space was designed to create a sense of individual and group identification with the space and with the company.

Thus, graphics and imaging were used to create identifiable "neighborhoods" throughout the building (see Figure 5.31).

**Figure 5.31** Workstation screens with images (Author)

To compensate for the possibility of other internal and external events that might influence employee responses, we compared the results from the Marketplace employees to a Control Group that did not receive a new workspace. With the control group, we were able to identify what changes occurred between groups that could be reasonably attributed to the new workspace, as opposed to other influences. Thus, in our analyses we were able to focus on changes that we can confidently say can be attributed to the work environment.

## C. Methods

We employed three methods of data collection: A questionnaire given to all employees who moved to the Marketplace and to a Control group, management interviews with selected department leads, and analyses of the time and cost of selected business processes.

## 1. Workplace Questionnaire

We developed and administered a questionnaire to all employees who moved into the Marketplace. Employees took the survey at two points in time: once prior to any changes and, subsequently, after employees had been in their new space for about two months. The Control group consisted of the same number of employees performing functionally similar work, and were all located at a nearby Headquarters building.

The questionnaire included 12 sections:

- Background information on the respondent
- Office environment evaluation
- Meeting space features
- Work group process
- Departmental collaboration and quality of departmental work output
- Communication
- Job satisfaction
- Job control
- Employee learning and growth
- Corporate image
- Space utilization
- Travel time and distances

Many of the individual questionnaire items developed had been pre-tested for reliability and successfully used in other research projects. Others were developed specifically for this project. The topics listed above were measured by statistically combining the individual survey questions into single indices that represent each topic. These indices were then used in the analysis.

## 2. Managers Interview

The purpose of these interviews was to identify the departmental business objectives that each department lead thought would be impacted by the new workspace. This information was ultimately used to relate changes in the workplace to improvements on the measures of these objectives. The data from these interviews was used to create the Strategic Business Model (see Figure 5.27) and used in evaluating the impacts of the project.

3. Business Process Analysis

An evaluation of selected business processes was conducted to assess the impact of the changes to the environment on individual and group business practices. Three business processes were evaluated within the Marketing department. The Marketing department was selected because many of their employees were affected by the move to the new building, and thus we inferred that the processes within that department might be impacted as well. In the analysis, we identified and mapped the process steps and collected metrics on cycle time and process quality, both before and after the move into the new space. The results for each process were compared to identify the differences in process quality and cycle time metrics, before and after the move to the Marketplace.

4. Design of the Study

In this study we were interested in examining the impacts of the consolidation, new work environment design, use of new furniture, and new space allocations on a variety of outcomes. The participants in this study were grouped into one of two treatment conditions:

1. Employees who moved into the Marketplace and experienced the consolidation and redesign (Experimental group)

2. Employees who did not move or experience the consolidation or receive new furniture (Control group)

The data was collected in two phases: Once before any change to the environment and before any move, and once after the move. Figure 5.32 shows the data collection points ("O" stands for "observation) and the timing of the move and consolidation (signified by the "X").

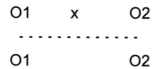

**Figure 5.32** Design of Study and Observations (Author)

## D. Results

### 1. Response Rates

Table 5.28 shows that a total of 1,264 surveys were distributed and 775 completed surveys were returned, for a 61 percent overall completion rate (17 percent to 30 percent is a typical return rate for survey studies).

| Table 5.28. Survey Response Rates Across Treatment Groups | | | | |
|---|---|---|---|---|
| | Market-place Pre-Change | Market-place Post-Change | Control Group Pre-Change | Control Group Post Change | Total |
| Surveys Distributed | 330 | 302 | 329 | 303 | 1264 |
| Surveys Returned | 252 | 150 | 214 | 159 | 775 |

### 2. Participant Characteristics

In this section we describe the general characteristics of the population we sampled, which includes gender, job description and departmental affiliation. These statistics are stated by percentage of the overall sample of 775 respondents. Table 5.29 shows a breakdown of all participants by gender. The table shows that gender is roughly evenly distributed across the sample, reflecting the makeup of the company at large.

Table 5.30 shows the breakdown of participants in this study by job

| Table 5.29 Participants by Gender | | |
|---|---|---|
| **Condition** | **% Female** | **% Male** |
| Experimental Group – Pre-Move | 58% | 42% |
| Control Group – Pre-Move | 40% | 60% |
| Experimental Group – Post-Move | 59% | 41% |
| Control Group – Post-Move | 41% | 59% |

category. The greatest proportion of participation is by professionals (58 percent) and work team leaders (19 percent), which is representative of the proportions of the actual population.

Table 5.31 shows a breakdown of participants by department. This table

| Table 5.30 Participants by Job Category | |
|---|---|
| **Job Description** | **Percentage of Total** |
| Executive | 2% |
| Work Team Leader | 19% |
| Professional Technical (accounting, engineer, project manager, etc.) | 58% |
| Transactional (clerk, data processing, customer service, etc.) | 9% |
| Administrative (support services, secretary) | 12% |

shows there is no one department that dominates the analysis, and that proportionally participation is representative of the actual population.

3. Variables Analyzed

| Table 5.31 Participants by Department | |
|---|---|
| **Department** | **Percentage of Total** |
| Finance/Legal/Tax | 6% |
| Sales and Distribution/A&D Accounts/Customer Centers/ Owned Stores | 8% |
| Product Development/Engineering | 2% |
| New Product Design/Research/Commercialization | 0.4% |
| HM for the Home | 2% |
| Marketing | 12% |
| eMarketing Communications/Creative Design | 4% |
| Information Architecture/Information Systems | 23% |
| HM Red | 2% |
| Operations/Logistics/Manufacturing/Engineering/Supply/ Quality | 18% |
| People Services | 9% |
| Workplace Strategy/Facilities | 3% |
| Customer Care | 1% |
| Corporate Training | 3% |
| HM Creative Office | 0.3% |
| Global Customer Solutions/ Global Partnership Accounts | 6% |
| Executive Administration | 0.3% |

In this section we summarize the results of our analyses for each of the areas of interest. The analyses are divided into these categories:

- Fit of design features to work needs within the workstation (lighting, storage, etc.)
- Overall layout of the space (support for work, identity)
- Individual and Group work process (collaboration, quality of output, interruptions, privacy, etc.)
- Degree to which the overall space supports collaboration

## 4. Method of Analysis

Participant responses to each variable analyzed in this study were compared on the two study conditions ("Experimental group," Marketplace employees that moved, and "Control group," employees that did not move) before and after the change. The overall statistical significance of each of these comparisons was computed using analysis of variance (ANOVA). In all cases the minimum criteria for statistical significance was set at $p <. 05$. The differences reported were statistically different, which means we can be sure at a 95 percent level of confidence that the percentage of change shown on each evaluation represents an actual change in participant's behavior or perceptions.

Table 5.32, and all other tables showing results information in this section, uses the same format for displaying the results. The column headings at the top of the table show the conditions "Pre-Move vs. Post-Move to Marketplace" and "Control Group." The name of each index or variable is shown on the leftmost column of the table. Immediately to the right of each variable name the percentage change is shown for the groups before and after the move. The "NS" indicates that there is no statistically significant difference between the groups being compared.

***Effects of Job Type.*** We conducted an analysis for each of the variables to determine if job type plays a role in employee responses to the study variables. The results showed no main effects for job type, meaning that employees across job types had very consistent responses to the majority of issues we investigated. Thus, for the analyses we present in this report, we report responses for all employees as a group and do not present analyses by job type. This has the positive side-effect of greatly simplifying both the presentation and interpretation of the results and their implications to the organization.

5. Effects of Workstation Design on Perceptions and Behaviors

Table 5.32 shows a summary of the impacts of the Marketplace consolidation and design project on employees' perceptions of workstation features. Our evaluation is compared against the baseline of employee assessments before any change. In Table 5.32 we summarize the findings for workstation design features, including visual and acoustic privacy, storage, lighting, workstation layout, comfort, adjustability of workstation, availability of alternative settings for individual work, air temperature, and personal identity. The results show that in no case did the perceptions of the control group change over the period of the two observations (see Table 5.32). This gives us confidence to suggest that the results of the changes to the experimental group were due to changes in the work environment, as opposed to being due to other external intervening conditions.

All percentage results listed are statistically significant. "NS" indicates there is no statistically significant difference between the groups being compared.

*Visual and Acoustic Privacy.* Employees who moved into the Marketplace reported a 9 percent decrease in the amount of visual and acoustical privacy. The results show there was no difference in visual and acoustical privacy between employees of the control group.

*Storage.* Employees who moved into the Marketplace reported a 15 percent decrease in the amount of sufficient storage capacity. The results show there was no difference between employees of the control group.

*Lighting.* Employees who moved into the Marketplace reported a 17 percent increase in the amount of lighting (both natural and task). The results show there was no difference between employees of the control group.

*Workstation Layout.* Workstation layout was defined through a number of measures, including work materials close at hand, arrangement of equipment and furnishings, work surface size, and ease of organizing work materials. Employees who moved into the Marketplace reported a 10 percent decrease in satisfaction with workstation layout. The results show there was no difference between employees of the control group.

*Comfort.* Employees who moved into the Marketplace reported no significant change in comfort and feel that the workstation layout can be used comfortably for long periods of time. The results show there was no difference between employees of the control group.

| Table 5.32 Percentage Change in Evaluation Scores on Workstation Design Features | | |
|---|---|---|
| **Changes to:** | **Comparison: Pre-Move vs. Post-Move to Marketplace – Experimental Group** | **Control Group** |
| Visual and acoustic privacy | -9% | NS |
| Storage | -15% | NS |
| Lighting | +17% | NS |
| Workstation layout | -10% | NS |
| Comfort | NS | NS |
| Adjustability of workstation | NS | NS |
| Availability of alternative settings to do individual work | +13% | NS |
| Air temperature | +15% | NS |
| Ability to express personal identity | -16% | NS |

***Adjustability of Workstation.*** Employees who moved into the Marketplace reported no significant change in the adjustability of workstations to fit their needs. The results show there was no difference between employees of the control group.

***Availability of Alternate Settings for Individual Work.*** Employees who moved into the Marketplace reported a 13 percent increase in the availability of alternative work settings for individual work. The results show there was no difference between employees of the control group.

***Air Temperature.*** Employees who moved into the Marketplace reported a 15% increase in their perceptions of comfort with air temperature. The results show there was no difference between employees of the control group.

***Personal Identity.*** Employees who moved into the Marketplace reported a 16 percent decrease in the ability to express personal identity through the display of personal items. The results show there was no difference between employees of the control group.

## 5. Effects of Workstation Design on Individual Work Process

In this section we describe the results of our analyses of the impact of the Marketplace on various aspects of individual work process, including sense of job control, type of communication methods used, communication between individuals, communication with management, interruptions/distractions, and the ability to handle confidential materials. Table 5.33 summarizes these findings. The results show that in no case did the perceptions of the control group change over the period of the two observations. This gives us confidence to suggest that the results of the changes to the experimental group were due to changes in the work environment, as opposed to being due to other external intervening conditions.

All percentage results listed are statistically significant. "NS" indicates

| Table 5.33 Summary of Changes to Individual Work Process | | |
| --- | --- | --- |
| Changes to: | Comparison: Pre-Move vs. Post-Move to Marketplace – Experimental Group | Control Group |
| Perception of control over job | NS | NS |
| Communication methods - Voice mail - Telephone - E-mail - Videoconference - Face-to-face - Paper memos | NS | NS |
| Communication between individuals | -7% | NS |
| Communication with management | -4% | NS |
| Interruptions and distractions | NS | NS |
| Ability to handle confidential material | -9% | NS |

there is no statistically significant difference between the groups being compared.

***Perceptions of Job Control.*** Employees who moved into the Marketplace reported no significant change in the sense of job control. Employees' perceptions remained the same about their control over their work schedules, making decisions, physical location in which work is performed, and the rate of speed in which work is done to meet objectives. The results also show there was no change in sense of job control across the two observations for the control group.

***Communication Methods.*** Employees who moved into the Marketplace reported no significant change in the type of communication methods they used. The use of face-to-face meetings did not increase after the move into the Marketplace. The results also show there was no change in communications methods used across the two observations for the control group.

***Communication between Individuals.*** Employees who moved into the Marketplace reported a 7 percent decrease in their communication between individuals. Communication was measured by employees' perceptions about the quality of communication with co-workers and the ability to have access to coworkers. The results also show there was no change in communications between individuals used across the two observations for the control group.

*Communication With Management.* Employees who moved into the Marketplace reported a 4 percent decrease in their communication with management. Employees have a lower perception about understanding clear future direction, expectations of performance and that their suggestions are valued and considered by management. The results also show there was no change in communications with management across the two observations for the control group.

*Interruptions and Distractions in the Workstation.* Employees who moved into the Marketplace have no significant change in the amount of interruptions or distractions in the workstation. This shows the importance of proper programming and design in reducing interruptions. The results also show there was no change in interruptions and distractions across the two observations for the control group.

*Ability to Handle Confidential Material.* Employees who moved into the Marketplace reported a 9 percent decrease in their ability to handle confidential materials. With the overall trend toward greater openness in the workspaces there appears to be a decreased ability to handle confidential materials within the workstation. The results also show there was no change in ability to handle confidential material across the two observations for the control group.

## 7. Effects of Workplace Design on Group Work Process

In this section we describe the results for various aspects of group work process, including work group process quality, communication between work groups, and the quality of departmental work output. Table 5.34 summarizes these findings. The results show that in no case did the perceptions of the control group change over the period of the two observations. This gives us confidence to suggest that the results of the changes to the experimental group were due to changes in the work environment, as opposed to being due to other external intervening conditions.

All percentage results listed are statistically significant. "NS" indicates there is no statistically significant difference between the groups being compared.

*Work Group Process Quality.* Employees who moved into the Marketplace had no significant change in their perception about the quality of their work teams' group processes. They continued to feel the same about being part of a team, how actively involved in team activities and decision-making

| Table 5.34 Effects of Workplace Design on Group Work Process | | |
|---|---|---|
| Changes to: | Comparison: Pre-Move vs. Post-Move to Marketplace – Experimental Group | Control Group |
| Work Group Process Quality<br>- Members feel like part of the team<br>- Members involved in team activities<br>- High participation<br>- Members support each other<br>- Members participate in decision-making | NS | NS |
| Communication between work groups | NS | NS |
| Quality of departmental work output | +3% | NS |

members are, and the support each individual receives from the other team members. The results also show there was no change across the two observations for the control group on this measure.

***Communication Between Work Groups.*** Employees who moved into the Marketplace had no significant change in their perception about their communication with other work groups. The results also show there was no change across the two observations for the control group on this measure.

***Quality of Departmental Work Output.*** Employees who moved into the Marketplace reported a 3 percent increase in their assessment of the quality of work products they receive from other departments. The results also show there was no change across the two observations for the control group on this measure.

8. Impact of Architectural Space on Support for Collaboration, Communicating Culture, and Travel

In this section we analyzed employees' perceptions of the fit between the large-scale design features associated with the consolidation and layout of the building and their work process needs. Table 5.35 summarizes our analysis of large-scale interior design and layout features, including spaces to support collaboration, building design and its support of departmental identity and a sense of belonging, communication of corporate identity, design impact on attraction of new employees, types of spaces used, and travel time and distances. The results show that in no case did the perceptions of the control group change over the period of the two observations. This gives us confidence to suggest that the results of the changes to the experimental group were due to changes in the work environment, as opposed to being due to other external intervening conditions.

All percentage results listed are statistically significant. "NS" indicates there is no statistically significant difference between the groups being compared.

*Spaces to Support Collaboration.* Employees who moved into the Marketplace reported a 16 percent increase in the availability of spaces for collaboration and support for large group meetings and project teams. They reported an increase in the availability of space for social activities and a general place for collaboration. The results also show there was no change across the two observations for the control group on this measure.

*Building Design Supports a Sense of Belonging.* Employees who moved into the Marketplace reported a 7 percent increase in the perception that the building conveys the shared values and image of the company to employees and that it supports their feelings of being part of the larger corporate community. The results also show there was no change across the two observations for the control group on this measure.

*Building Design Supports Departmental Identity.* Employees who moved into the Marketplace reported an 11 percent increase in the ability of the workspace to convey their department's identity. The results also show there was no change across the two observations for the control group on this measure.

*Building Design Communicates Corporate Identity to Customers.* Employees who moved into the Marketplace reported a 31 percent increase in the perception that the building conveys the company's corporate values to their customers. The results also show there was no change across the two observations for the control group on this measure.

*Building Design Will Attract New Employees.* Employees who moved into the Marketplace reported a 36 percent increase in the perception that the workspace will help attract new talented workers. The results also show there was no change across the two observations for the control group on this measure.

*Types of Spaces Utilized by Employees.* Employees who moved into the Marketplace reported no significant changes in where they spend their time working. As Table 5.35 indicates, there are slight increases in the time spent working in the building in which the employee's workstation is located. It also shows that collaborative spaces, such as meeting rooms and public areas, are used slightly more often than before the move. The results also show there was no change across the two observations for the control group on this measure.

| Table 5.35 Impact of Architectural Space on Support for Collaboration, Communicating Culture, and Travel | | |
|---|---|---|
| Changes to: | Comparison: Pre-Move vs. Post-Move – Experimental Group | Control Group |
| Spaces to support collaboration<br>- Availability of spaces for collaboration<br>- Support for large group meetings<br>- Support for project teams<br>- Social activities<br>- General support for collaboration | +16% | NS |
| Building design supports sense of "belonging" in employees | +7% | NS |
| Design supports departmental identity | +11% | NS |
| Design communicates corporate identity to customers | +31% | NS |
| Design will attract new employees | +36% | NS |
| Types of spaces employees spend time in: | | |
| - unassigned workstation in your building | 0.95% – 3.91% | NS |
| - assigned workstation in your building | 15.23 – 14.39% | NS |
| - in meeting rooms in your building | 5.67 – 5.04% (NS) | NS |
| - in other public areas of your building | 2.58 – 3.0% (NS) | NS |
| - campsite, not in your building | 1.64 – 1.09% | NS |
| - meeting room, not in your building | 2.95 – 2.36% | NS |
| - home office | 3.76 – 3.55% (NS) | NS |
| - other non-work location | 1.47 – 1.42% (NS) | NS |
| Travel time:<br>Average travel time per week | 44.58 – 44.50 (NS) | NS |
| Travel distance:<br>Average miles per week | 23.69 – 15.02 | NS |

***Travel Time.*** Employees who moved into the Marketplace reported no significant changes in the amount of time traveling between buildings when compared to travel times before the consolidation.

***Travel Distance.*** Employees who moved into the Marketplace reported no significant changes in the distances traveled between buildings when compared to travel distances before the consolidation.

## 9. Regression Analyses

A Regression Model is a statistical model showing the predictive relationships between a group of variables, such as design features, and an outcome variable, which can be a measure of a behavior or work process. We created four models to show the work environment and organizational variables that predict key outcomes of this study. These models include outcome measures of quality of departmental work products, job satisfaction, job control, and quality of internal group process. The purpose of creating these models is to create strategic information about work process and design feature issues that can be generalized to all areas of an organization.

***Regression Model: Quality of Departmental Work Products.*** We created and tested a model to assess the relationship between design and organization or management variables, and the assessed quality of departmental work outputs or products. This outcome measure, quality of departmental work products, was measured as an index of satisfaction with the quality of the deliverables from all 22 departments involved in this study. All participants in this study provided an assessment of the quality of deliverables or outputs for each department on a 5-point scale, ranging from low to high quality. Each group of 22 assessments was combined into a single index, with an internal reliability of $\alpha = 0.82$. Data used in this analysis was collected both prior to and after the move to the Marketplace building.

A variety of models were created and tested. The most powerful model revealed that the assessed quality of the work products of departments is predicted by two variables: Availability of collaborative space, and effective communication from management to employees (see Figure in Table 5.36). These two variables predict 10 percent of the variance in assessed quality of departmental work products ($R^2 = 0.10$) (see Table 5.36). Both of the predictor variables contribute about equally to the prediction of the outcome measure (spaces to support collaboration = beta 0.19, and communication from management = beta 0.22).

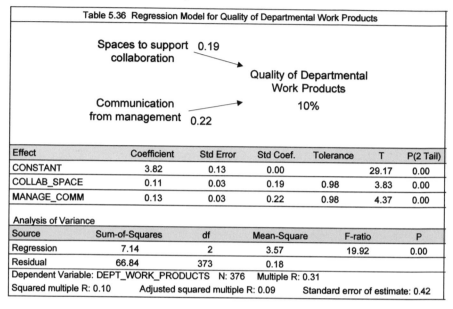

| Effect | Coefficient | Std Error | Std Coef. | Tolerance | T | P(2 Tail) |
|---|---|---|---|---|---|---|
| CONSTANT | 3.82 | 0.13 | 0.00 | | 29.17 | 0.00 |
| COLLAB_SPACE | 0.11 | 0.03 | 0.19 | 0.98 | 3.83 | 0.00 |
| MANAGE_COMM | 0.13 | 0.03 | 0.22 | 0.98 | 4.37 | 0.00 |

Analysis of Variance

| Source | Sum-of-Squares | df | Mean-Square | F-ratio | P |
|---|---|---|---|---|---|
| Regression | 7.14 | 2 | 3.57 | 19.92 | 0.00 |
| Residual | 66.84 | 373 | 0.18 | | |

Dependent Variable: DEPT_WORK_PRODUCTS   N: 376   Multiple R: 0.31
Squared multiple R: 0.10        Adjusted squared multiple R: 0.09        Standard error of estimate: 0.42

The first predictor variable, spaces to support collaboration, was created as an index of seven survey items, including availability of spaces for small meetings, spaces for large group meetings, dedicated spaces for project teams, spaces for social activities, and technology and tools in meeting spaces. The reliability of this scale was $\alpha = 0.82$.

The second predictor variable, communication from management to employees, was created as an index of four survey items, including the degree to which senior management gives a clear picture of where the company is headed, senior management paying attention to ideas and suggestions from people, departments having adequate time to prepare for changes, and manager communication of work expectations to employees. The reliability of this scale was $\alpha = 0.93$.

***Interpretation.*** This model shows how physical design features and organizational/management issues can combine to influence important organizational outcomes such as the quality of the work products at the departmental level.

This model also suggests that if communication from management can be enhanced, and the availability and variety of meeting spaces be expanded or further integrated as a design strategy, significant, positive effects on the ongoing quality improvement may be realized at the departmental or business unit level. While the 10 percent impact is a relatively small amount, it is still significant because it shows us the environmental features that can be manipulated to affect collaboration and, ultimately, effectiveness.

***Regression Model: Job Satisfaction.*** We created and tested a model to assess the relationship between design and organization or management variables and employee job satisfaction. This outcome measure, job satisfaction, was measured as an index of three questions assessing satisfaction with current job, recommend this employer to a friend, and degree to which the company cares about its employees. The reliability of this scale was tested, resulting in an internal reliability of $\alpha = 0.78$. This is a well-tested index of job satisfaction that has been used in numerous other studies. Data used in this analysis was collected both prior to and after the move to the Marketplace.

A variety of models were created and tested. The most powerful model, which predicts a full 47 percent of the variance in job satisfaction scores, revealed that job satisfaction was predicted by a combination of five variables related to communication, control over the job, and work environment design (see Table 5.37).

These predictor variables will be discussed in descending order of the magnitude of their impact on the outcome variable. The first predictor variable, communication from management to employees, was created as an index of four survey items, including the degree to which senior management gives a clear picture of where the company is headed, senior management paying attention to ideas and suggestions to people, departments having adequate time to prepare for changes, and manager communication of work expectations to employees. The reliability of this scale was $\alpha = 0.93$.

The second predictor variable, job control, was created as an index of four survey items, including control over work schedule, decisions over how to do the job, control over physical location of work, and control over speed at which work must be accomplished. The reliability of this scale was $\alpha = 0.77$.

The third variable, layout of interior workstation space, is an index of five survey items, including paper-based materials close at hand, arrangement of furnishings and equipment supports work, size of work surfaces appropriate, ease of organizing work materials, and sufficient filing.

The fourth variable, individual communication, was created as an index of three survey items, including quality of individual communication with co-workers, ease of accessing co-workers, ease of individual accessibility to co-workers.

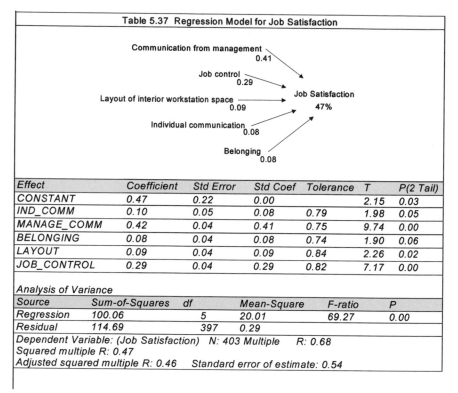

| Effect | Coefficient | Std Error | Std Coef | Tolerance | T | P(2 Tail) |
|---|---|---|---|---|---|---|
| CONSTANT | 0.47 | 0.22 | 0.00 | | 2.15 | 0.03 |
| IND_COMM | 0.10 | 0.05 | 0.08 | 0.79 | 1.98 | 0.05 |
| MANAGE_COMM | 0.42 | 0.04 | 0.41 | 0.75 | 9.74 | 0.00 |
| BELONGING | 0.08 | 0.04 | 0.08 | 0.74 | 1.90 | 0.06 |
| LAYOUT | 0.09 | 0.04 | 0.09 | 0.84 | 2.26 | 0.02 |
| JOB_CONTROL | 0.29 | 0.04 | 0.29 | 0.82 | 7.17 | 0.00 |

Analysis of Variance

| Source | Sum-of-Squares | df | Mean-Square | F-ratio | P |
|---|---|---|---|---|---|
| Regression | 100.06 | 5 | 20.01 | 69.27 | 0.00 |
| Residual | 114.69 | 397 | 0.29 | | |

Dependent Variable: (Job Satisfaction)  N: 403 Multiple    R: 0.68
Squared multiple R: 0.47
Adjusted squared multiple R: 0.46    Standard error of estimate: 0.54

The fifth variable, belonging, is an index of five survey items, including workspace design that supports feelings of ownership, belonging to the work group and to the company, feelings of emotional "comfort," and the building conveys shared values and image of the company and promotes a sense of community.

The two most significant predictor variables, management communicates to employees (beta = 0.41), and job control (beta = 0.29), contribute about equally to the prediction of job satisfaction. The remaining variables have less, but still significant, predictive strength.

***Interpretation.*** This model shows that a combination of human/behavioral (communication; job control) and workplace attributes (fit of workspace layout; design fosters a sense of belonging) greatly influence employee job satisfaction, predicting 47 percent of the variance in job satisfaction scores (to put this percentage in perspective, studies of this nature typically report predictive models of 8 percent to 15 percent). Other analyses revealed that this model is valid regardless of job level in the organization or gender. These analyses show this model is generalizable across the organization, applying to all employees regardless of job type or gender.

Thus, a design strategy that supports individual work (layout) and sense of group identity (belonging), along with management practices that promote communication between employees and management and control over the job for employees, will result in greater job satisfaction. Companies regularly use measures of job satisfaction as leading indicators of employee retention trends.

***Regression Model: Job Control.*** We created and tested a model to assess the relationship between design and organization or management variables, and employees' sense of control over their job (see Table 5.38). In the literature on work and health, a direct link has been found between degree of control over the job and stress and health issues, such as coronary heart disease (CHD). Thus, it is important to optimize the design of the workspace to enhance job control, if possible. In this model, we found that a mix of organizational and workplace design attributes predicted job control.

The construct of job control was measured as an index of four items assessing control over work schedule, decisions affecting the job, physical location of work, and rate of speed of work. This is a well-tested index of group process that has been used in numerous other studies. The internal reliability of this index is $\alpha = 0.77$.

The model we arrived at predicts 12 percent of the variance in job control scores, and showed that job control was predicted by a combination of four variables related to aspects of communication and work environment design (see Table 5.38), including: job type, interruptions, communication between individuals, and space support for individual work. These predictor variables will be discussed in descending order of the magnitude of their impact on the outcome variable.

The first predictive variable, job type (beta = -0.19), is an indirect measure of rank within the organization. The model shows that the higher the rank in the organization, the greater the control over the job.

The second predictive variable, interruptions (beta = -0.17), is an index of two items assessing the level of interruptions, and distractions from background noise while in the workspace. The model shows that greater interruptions and distractions predict a lower sense of job control.

The third variable, individual communication (beta = 0.16), was created as an index of three survey items, including quality of individual communication with co-workers, ease of accessing co-workers, ease of individual accessibility to co-workers. The model shows that increased communication between individuals is related to increased job control.

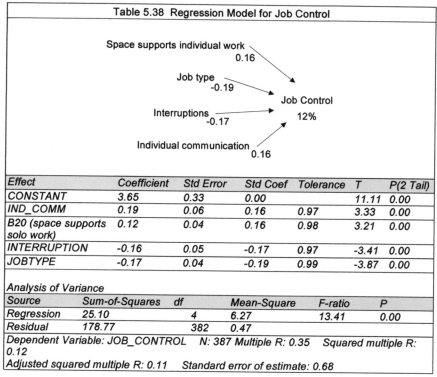

Table 5.38  Regression Model for Job Control

| Effect | Coefficient | Std Error | Std Coef | Tolerance | T | P(2 Tail) |
|---|---|---|---|---|---|---|
| CONSTANT | 3.65 | 0.33 | 0.00 | | 11.11 | 0.00 |
| IND_COMM | 0.19 | 0.06 | 0.16 | 0.97 | 3.33 | 0.00 |
| B20 (space supports solo work) | 0.12 | 0.04 | 0.16 | 0.98 | 3.21 | 0.00 |
| INTERRUPTION | -0.16 | 0.05 | -0.17 | 0.97 | -3.41 | 0.00 |
| JOBTYPE | -0.17 | 0.04 | -0.19 | 0.99 | -3.87 | 0.00 |

Analysis of Variance

| Source | Sum-of-Squares | df | Mean-Square | F-ratio | P |
|---|---|---|---|---|---|
| Regression | 25.10 | 4 | 6.27 | 13.41 | 0.00 |
| Residual | 178.77 | 382 | 0.47 | | |

Dependent Variable: JOB_CONTROL   N: 387 Multiple R: 0.35   Squared multiple R: 0.12

Adjusted squared multiple R: 0.11    Standard error of estimate: 0.68

The fourth variable, space supports individual work (beta = 0.16), is a measure of satisfaction with the availability of different workspace settings in which to perform individual work. The model shows that greater availability of spaces to perform individual work is a predictor of control over the job.

*Interpretation.* While workspace cannot do anything to increase employee rank within the company, the work environment can influence the other three predictive variables. The theme here appears to be creating a balance that supports communication between individuals while having design characteristics that minimize interruptions and distractions. In addition, having an interior layout that provides a variety of spaces that support individual work is critical to providing a sense of control over the job.

Clearly the tension between the need for communication between individuals and workspaces that support non-distracted individual work needs to be addressed in order to drive a workplace strategy that supports job control. Because of the link between job control and health issues, it is important to optimize the design of the workspace to enhance job control, where possible.

***Regression Model: Quality of Group Process.*** We created and tested a model to assess the relationship between design and organization or management variables, and quality of group process (see Table 5.39). In the management research literature, group process measures are often used to assess group effectiveness. Quality of group process was measured as an index of five items assessing sense of group membership, involvement and participation in activities including decision-making, and member supportiveness towards each other. The reliability of this scale was tested, resulting in an internal reliability of $\alpha = 0.90$. This is a well-tested index of group process that has been used in numerous other studies.

The model we arrived at predicts 31 percent of the variance in quality of group process scores, and showed that group process was predicted by a combination of four variables related to aspects of communication and work environment design (see Table 5.39). These predictor variables will be discussed in descending order of the magnitude of their impact on the outcome variable.

The first predictor variable, communication from management to employees, was created as an index of four survey items, including degree to which senior management gives a clear picture of where the company is headed, senior management paying attention to ideas and suggestions from people, departments having adequate time to prepare for changes, and manager communication of work expectations to employees. The reliability of this scale was $\alpha = 0.93$.

The second predictor variable, communication between work teams, was measured through a single question item.

The next two predictor variables are related to work space design. Although they contribute relatively less to the prediction of the outcome measure, their contribution is still important to the model and statistically significant.

The third predictor variable, spaces to support collaboration, was created as an index of seven survey items, including availability of spaces for small meetings, spaces for large group meetings, dedicated spaces for project teams, spaces for social activities, and technology and tools in meeting spaces. The reliability of this scale was $\alpha = 0.82$. The second predictor variable, communication from management to employees, was created as an index of four survey items, including degree to which senior management gives a clear picture of where the company is headed, senior management paying attention to ideas and suggestions from people, departments having adequate time to prepare for changes, and manager communication of work expectations to employees. The reliability of this scale was $\alpha = 0.93$. The fourth predictor variable, workstation adjustability, was measured through a single survey item measuring the ability to adjust workstation to fit work needs.

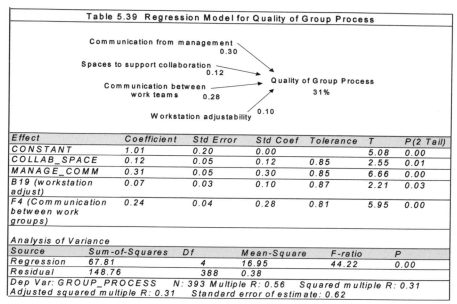

Table 5.39 Regression Model for Quality of Group Process

| Effect | Coefficient | Std Error | Std Coef | Tolerance | T | P(2 Tail) |
|---|---|---|---|---|---|---|
| CONSTANT | 1.01 | 0.20 | 0.00 | | 5.08 | 0.00 |
| COLLAB_SPACE | 0.12 | 0.05 | 0.12 | 0.85 | 2.55 | 0.01 |
| MANAGE_COMM | 0.31 | 0.05 | 0.30 | 0.85 | 6.66 | 0.00 |
| B19 (workstation adjust) | 0.07 | 0.03 | 0.10 | 0.87 | 2.21 | 0.03 |
| F4 (Communication between work groups) | 0.24 | 0.04 | 0.28 | 0.81 | 5.95 | 0.00 |

Analysis of Variance

| Source | Sum-of-Squares | Df | Mean-Square | F-ratio | P |
|---|---|---|---|---|---|
| Regression | 67.81 | 4 | 16.95 | 44.22 | 0.00 |
| Residual | 148.76 | 388 | 0.38 | | |

Dep Var: GROUP_PROCESS    N: 393 Multiple R: 0.56    Squared multiple R: 0.31
Adjusted squared multiple R: 0.31    Standard error of estimate: 0.62

The two most significant predictor variables, management communicates to employees (beta = 0.30), and communication between work teams (beta = 0.28), contribute about equally to the prediction of group process. The remaining work environment design variables, availability of collaborative space (beta = 0.12), and workstation adjustability (beta = 0.10), have less, but still significant, predictive strength.

*Interpretation.* This model shows that a combination of communication issues (communication between management and employees, and between work teams) and workplace attributes (availability of collaborative space, and workstation adjustability) greatly influence quality of group process, predicting 31 percent of the variance in this outcome scores. (To put this percentage in perspective, studies of this nature typically report predictive models of 8 percent to 15 percent.) Other analyses revealed this model is valid regardless of job level in the organization or gender. These analyses show this model is generalizable across the organization, applying to all employees regardless of job type or gender.

Clearly the issues of communication and flexible and collaborative work-spaces are conceptually related. In addition, these variables can be logically related to the quality of group process. Thus, a design strategy intended to enhance group process needs to incorporate spaces that support the flow of collaborative group work in a flexible manner. The same concept applies at the individual level with the workstation needing to accommodate the work-flow transitions from individual to group activities and back again. The availability of collaborative spaces and management practices that support communication are physical and organizational characteristics that reinforce each other to enhance group process.

As is the case with the other models, we find that a facility design strategy should consider, or even be explicitly combined with, existing aspects of management practices, such as the desire to facilitate communication throughout the organization. This model provides empirical support for the common sense notion that the provision of flexible spaces to support collaboration will enhance group process.

## 10. Summary of Regression Models.

In this section, we tested models related to outcomes of quality of departmental work products, quality of group process, job satisfaction, and job control. The models show that having a mix of spaces that support group and individual work will directly influence quality of group process, quality of the outputs of departments, and sense of job control at the individual level. This research also suggests that a facility design strategy that supports individual and collaborative work will result in positive benefits related to improved work performance at the departmental unit, group, and individual levels.

In addition, these models also revealed the complex nature of the work system, which must incorporate an understanding of organizational and management behaviors, as well as building and interior design features. A facility design strategy must embrace and leverage organizational-level issues such as management communication to support positive outcomes related to work process, satisfaction, and control in employees.

## 11. Business Process Analysis

To determine whether business process improvements resulted from the move to the new Marketplace facility, we collected business process efficiency data from several internal groups who moved into the Marketplace. We created process maps to measure the amount of the time, resources and overall elapsed cycle time related to repeatable business processes prior to the move to the Marketplace and four months after the move. We used the data to determine whether there were any changes in process efficiency after the move the Marketplace.

*Processes Analyzed.* In this study we conducted analyses of three ongoing business processes within the company. Within the Marketing department we mapped the Press Release Process. Within the Information Architecture department, we mapped the New Product Specification Process. Within the eMarketing Department we mapped the Trade Show Promotional Web Site

Process. These departments and processes were selected because they all participated in the move from several dispersed locations. We felt that measurable changes in process quality could result from the move to the Marketplace building because of the closer proximity of members of the departments to each other, and to their internal customers.

***Background on Process Teams.*** The sizes of the teams supporting the three processes that we mapped varied in size from two to eight team members. Prior to moving into the Marketplace, members of each process team had resided in another location. Each of the teams worked with other groups outside the original location, as well as outside vendors. Although there was some collaboration between work groups to complete their business process, for the most part each team worked independent of one another before the move.

***Process Metrics.*** To collect the process metrics used in these analyses, we first mapped the existing process with each group prior to the move. Several months after the groups had relocated in the Marketplace we reviewed the original process map with each group and updated the metrics data. Thus, we collected pre-move and post-move cycle time measures for each of the three processes. We found that the processes themselves did not change, after the move.

Specifically, we gathered each of the following items of information pertaining to each process:

♦ The amount of time required to complete a single process step
♦ The resources (i.e., internal and external groups, outside vendors) involved in the completion and decision making required for each process step
♦ The technology and tools used to complete each process step
♦ The overall amount of actual process time, and overall elapsed time required to complete an entire process cycle

***Rating Quality of Process Outputs.*** Additionally, we asked each team responsible for the process to rate the overall quality of the outputs of the process on a five-point Likert-type scale, ranging from low to high quality (1 = poor, 2 = fair, 3 = average, 4 = good, and 5 = excellent). General comments and observations from team members were also captured to convey the process teams perceptions of how the new work environment affected their work.

12. Business Process Analysis Results

***Trade Show Promotional Web Site Process.*** Data was gathered for five phases of this process, including Describe, Define, Design, Production, Test and Launch. The time measurements we collected for this process show there was no significant change in the cycle time required to complete the process after the move to the Marketplace (see Table 5.40).

*Data Analysis and Observations.* Although there was little measured change to cycle time after the move, it is possible that additional measurements following a longer duration at the Marketplace could reveal such efficiency improvements. Based on the initial move to the Marketplace, team members reported that they felt more collaborative as a team and their work processes felt more efficient.

A decrease in travel time (as well as the elimination of an expense payment by the organization) was realized based on the elimination of travel between locations as a part of this process. The team felt the overall quality of team member relationships increased through the ability to have impromptu, face-to-face meetings at team member workstations. The collaborative workspace for the team allowed managers to communicate process needs and work requirements in person to help foster better team relations.

*New Product Specification Process.* Data was gathered for three phases of this process, including New Product Identification to Product Name/Coding, New Product Visual/Pricing to PIW Reporting, and New Product Extracting to Posting.

*Data Analysis and Observations.* After the move into the Marketplace there was a 7.5 percent reduction in the overall process time required to complete the New Product Specification Process (see Table 5.40). We feel that the reduction in process step time might be related to the improved adjacencies of the team members. Before the move to the Marketplace, the team members were at different locations, and communication was based on E-mail or phone calls. This made the process to collect and verify product information difficult, increasing cycle time, wait times, and affecting product information accuracy. Now that the teams are located at the Marketplace they are able to communicate face-to-face and on an "as-needed basis," enhancing process effectiveness and performance.

*Press Release Process.* Data was gathered for three phases of the "Press Release" business process, including Checklist Completion to Press Release Development, Press Release Data Validation to Draft Press Release, and Approved Press Release to Distribution.

*Data Analysis and Observations.* The measurements collected for the Press Release Process revealed there was no significant change in cycle time required to complete the process after the move to the Marketplace (see Table 5.40).

Since the teams they collaborate with as a part of this process are now located at the Marketplace, it's easier for members of the Press Release team to locate other team members to obtain immediate feedback and to have impromptu discussions.

| Table 5.40 Analysis of Pre-Move and Post-Move Business Process Metrics | | |
|---|---|---|
| *Marketing Department – Press Release Process* | | |
| | **Pre-Move** | **Post-Move** |
| **Cycle Time in Hours** | 193.4 | 193.4 |
| *Information Architecture Department – New Product Specification Process* | | |
| | **Pre-Move** | **Post-Move** |
| **Cycle Time in Hours** | 155.2 | 147.4 |
| *eMarketing Department – Trade Show Promotional Website Process* | | |
| | **Pre-Move** | **Post-Move** |
| **Cycle Time in Hours** | 287 | 289 |

13. Business Process Productivity Analysis

Increased productivity is generally defined as a reduction in cost (time) to produce a given unit of work output. As previously discussed, we found time reductions in the New Product Specification process (see Table 5.40). To further measure the impact of this time reduction on costs, we determined the number of times per calendar year this process was conducted and the number of team members involved in the process. This information was used to calculate the financial impact of these improvements.

The 7.5 percent decrease in cycle time for the New Product Specification process contributes $375 of cost savings to the business each time the process occurs. Based on the number of process occurrences, $4,000 of cost savings are realized annually for just this one process. These improvements in dollar amounts are based on *time* and average market *salary* data, for an assessment of productivity. These results could be extrapolated for other business processes within the business, that we did not measure.

The following assumptions have been made in determining the impact of the improvements

♦ Labor costs based on average market salary in that area of the country at the time of this study, for professional office workers.

♦ Process cycle time was collected during process mapping sessions and represent estimates of time for completion of each process before and after the consolidation by the employees conducting the work.

♦ Groups delivering the process estimated number of process completions per year.

***Results: Quality Evaluation of Departmental Work Products.*** We asked individual members of all 22 departments to provide an assessment of the quality of work products provided by every other department for which they had direct knowledge of these results. This outcome measure, quality of departmental work products, was measured as an index of satisfaction with the quality of the deliverables from all 22 departments involved in this study. All participants in this study provided an assessment of the quality of deliverables or outputs for each department on a five-point scale, ranging from low to high quality. Each group of 22 assessments was combined into a single index, with an internal reliability of $\alpha = 0.82$.

We used these assessments to provide another dimension to the performance measures used in this study. Table 5.41 shows a summary of these independent evaluations of departmental performance for the three departments for which we conducted the Business Process Analyses.

| Table 5.41 Evaluation of Quality of Departmental Work Products Pre/Post Move | | | |
|---|---|---|---|
| **Department** | **Pre-Move (Mean)** | **Post-Move (Mean)** | **Significance** |
| Marketing | 4.71 | 4.86 | P <.001 |
| Information Architecture | 4.43 | 4.74 | P <.001 |
| eMarketing | 4.56 | 4.72 | P <.001 |

As Table 5.41 shows, the evaluations of quality of work output improved significantly after the move to the Marketplace for all three departments.

14. A Model Linking Workplace Design to Business Process Effectiveness

A "regression model" is a statistical model showing the predictive relationships between a group of variables (such as design features) and an outcome variable (which can be a measure of a behavior or work process). We created a model from the results of the New Product Specification business process analysis. The process cycle time measure was used as the dependent variable (the outcome measure - the item being "predicted"). The intent of this model was to show the predictive relationship between design features of the workspace and changes in objective measures of business process effectiveness.

The model we developed revealed that process cycle time for the New Product Specification Process is predicted by six variables (see Figure 5.42):

Availability of collaborative space (beta = 0.31), lighting (beta = 0.33), quality of group process (beta = 0.23), design of interior space supports shift from individual to collaborative work (beta = 0.35), amount of time spent in unassigned workspace (beta = 0.15), and quality of storage in the workstation (beta = 0.23) (see Table 5.42).

These six variables predict 32 percent of the variance in cycle time ($R^2$ = 0.32) (see Table 5.42). A model with an explanatory magnitude of 32% for a study of this type is a very powerful model. The Analysis of Variance calculated shows that the model is statistically significant ($F$ = 4.01, df = 6,52, p<.01) (see Table 5.42).

The beta values for each variable in the model indicate the relative contribution of each variable to the outcome measure. The values represent the "one way" or unidirectional correlations between the predictor and outcome variables (see "Std. Coef" column, Table 5.42). The predictor variables are discussed in descending order of the magnitude of their beta values.

♦ The first predictor variable, "design of interior space supports employee ability to shift from individual work to collaboration," was measured by one questionnaire item.

♦ The second predictor variable, lighting, was created by combining two items, including overall lighting (overhead, task, daylight) at the workstation is appropriate, and sufficient amount of natural light at workspace.

♦ The third predictor variable, spaces to support collaboration, was created as an index of seven survey items, including availability of spaces for small meetings, spaces for large group meetings, dedicated spaces for project teams, spaces for social activities, and technology and tools in meeting spaces. The reliability of this scale was $\alpha$ = 0.82.

♦ The fourth predictor variable, group process quality, was created as an index of five survey items, including members feel like part of the group, all members involved in group activities, high participation of members, members supportive of each other, members participate in decision-making. The reliability of this scale was $\alpha$ = 0.90.

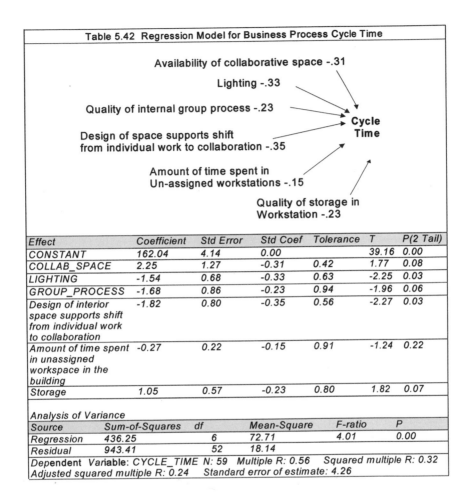

Table 5.42 Regression Model for Business Process Cycle Time

Availability of collaborative space -.31

Lighting -.33

Quality of internal group process -.23

Design of space supports shift
from individual work to collaboration -.35

Cycle
Time

Amount of time spent in
Un-assigned workstations -.15

Quality of storage in
Workstation -.23

| Effect | Coefficient | Std Error | Std Coef | Tolerance | T | P(2 Tail) |
|---|---|---|---|---|---|---|
| CONSTANT | 162.04 | 4.14 | 0.00 | | 39.16 | 0.00 |
| COLLAB_SPACE | 2.25 | 1.27 | -0.31 | 0.42 | 1.77 | 0.08 |
| LIGHTING | -1.54 | 0.68 | -0.33 | 0.63 | -2.25 | 0.03 |
| GROUP_PROCESS | -1.68 | 0.86 | -0.23 | 0.94 | -1.96 | 0.06 |
| Design of interior space supports shift from individual work to collaboration | -1.82 | 0.80 | -0.35 | 0.56 | -2.27 | 0.03 |
| Amount of time spent in unassigned workspace in the building | -0.27 | 0.22 | -0.15 | 0.91 | -1.24 | 0.22 |
| Storage | 1.05 | 0.57 | -0.23 | 0.80 | 1.82 | 0.07 |

Analysis of Variance

| Source | Sum-of-Squares | df | Mean-Square | F-ratio | P |
|---|---|---|---|---|---|
| Regression | 436.25 | 6 | 72.71 | 4.01 | 0.00 |
| Residual | 943.41 | 52 | 18.14 | | |

Dependent Variable: CYCLE_TIME N: 59 Multiple R: 0.56 Squared multiple R: 0.32
Adjusted squared multiple R: 0.24 Standard error of estimate: 4.26

♦ The fifth predictor variable, storage, is represented by a single survey item, sufficient filing space in the workstation.

♦ The sixth predictor variable, amount of time spent in unassigned workspace in the building is represented by a single survey question.

*Implications of the Model.* This model has predictor variables that represent primarily work environment design features designed to support flexible, group-oriented work. Only one variable, quality of group process, is not a measure of the work environment. Quality of collaborative space, lighting, quality of internal group process, design of interior space to support shift of work from individual to collaborative, amount of time spent in unassigned workstations, and storage, are all inversely, or negatively, related to length of process cycle time (see Table 5.42).

In other words, the higher the ratings on these characteristics, the greater the reduction in process cycle time.

***Analysis of Output and Cycle Time Measures for the eMarketing Department.*** In addition to this path model, which for the first time shows how specific workspace design features impact quantitative measures of business process efficiency, we have evidence gathered from the survey in which employees rated the quality of departmental work output. As discussed in an earlier section, the quality of work products of the eMarketing department did rise significantly after the move to the Marketplace - at least according to the ratings of independent observers.

In Figure 5.33, we have plotted process cycle time (for the New Product Specification Process) against the independent assessment of the quality of departmental output for the eMarketing department (which is responsible for the New Product Specification Process). Figure 5.33 also shows an interaction between the two measures in which process cycle time decreases in the time period after the move to the Marketplace, and the independent assessment of work product quality rises.

These findings would be exciting if we had only found that design features reduced the cost of a given process. What this research has shown is that, due to improved workplace design features, process costs been reduced, and in addition, the quality of the output (beyond the time to produce the product) has been increased. This, in effect, is a simple example of a classic definition of productivity. Quality of output rose while the time (cost) to deliver the output was reduced.

15. Conclusions

These results show the importance of examining work effectiveness from not only a time/cost perspective (as we did with the process metrics) but also from the perspective of the quality of the output. In a previous section, we constructed a regression model showing that spaces to support collaboration, and communication from management were the two predictors of the quality of departmental work products.

Thus we are able to show that both time and quality measures improved after the move to the Marketplace, and specific features of the design of the space at least in part, predicted both of these important outcomes.

The results of this study show that the strategic move made by the organization, from a work environment in which employees were scattered across different locations, to an environment in which departments are closely accessible with the use of an open office floor plan, has resulted in improvements in communication, collaboration, and business process efficiency and effectiveness.

Considering that the majority of individuals moved from offices with relatively greater enclosure to a more open environment, the finding that overall privacy perceptions decreased is not surprising. This finding suggests that the layout of the consolidation and new workstations are for the most part serving the work process needs of the employees, and that employees are adjusting to the openness of their environment. Employees indicated they are having less difficulty getting in touch with others, collaboration is improving, and departments are responding more quickly to each other. The findings suggest these improvements are a result of improved adjacencies, layout and the consolidation. Prior to the move to the Marketplace, individuals had to travel great distances to collaborate. In the new environment, despite (or perhaps as a result of) less privacy and more interruptions, employees can more easily contact and collaborate with others. Thus, the positive side of increased interruptions is more effective collaboration. Finally, it also may be that employees are less concerned than expected about their personal privacy because of the positive impact they are seeing the open environment have on group communication and workflow.

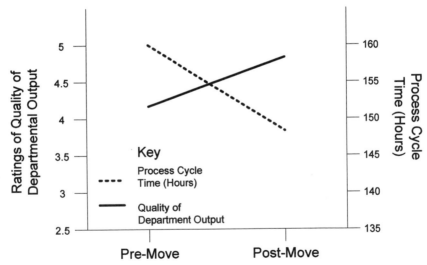

**Figure 5.33** Output Quality and Cycle Time Measures for eMarketing Department: New Product Specification Process (Author)

Based on this team's experience from previous research and a review of the literature, the results of this study are similar to what we've found in terms of behavioral impacts and work process improvements. It is important to reiterate that we believe the positive benefits that we found are due in part to the redesign of spaces and layout (of which the new furniture played a role) and partly due to the consolidation. Consolidation does not provide much benefit if it is not paired with correct adjacencies between people and departments to facilitate the flow of work between people and business processes between departments.

## VIII. CASE STUDY: COMPARISON OF TWO CALL CENTERS USING DIFFERENT WORKSPACE MODELS ON AGENT PERFORMANCE AND HEALTH COSTS

### A. Background

This telecommunications company is based in the US. The company is continually in the process of acquiring new Real Estate assets, with many in the form of call centers. These call centers are not standardized on any one facility/workspace model. The intent of this study was to compare two call centers that use different workspace models, but are organizationally similar (in terms of operational purpose, management, job design, and technology). While there was interest in comparing the performance of the two call centers directly, the strategic intent was to combine the data from the two locations to identify broad facility and workstation design features that are related to individual and organizational effectiveness.

### B. Purpose of the Project

The purpose of the project was to compare the effectiveness of two different work environment models; one using a frame-and-tile workstation furniture system, and the other using a pole-based furniture system. An additional intent was to combine the data collected from both locations to investigate the relationship between design features and business performance outcomes for call center agent jobs, and for call centers in general.

### C. The Locations

Both call centers are located within two hundred miles of each other in the Midwest area of the U.S. Both centers were built within two years of each other. The Illinois call center houses about 1200 employees, primarily call center agents. These agents are seated within frame-and-tile workstations clustered together in small groups (see Figure 5.34).

**Figure 5.34** View of agent workstations at Illinois facility (Author)

Managers are seated in larger workstations against the perimeter walls of the facility (see Figure 5.35).

**Figure 5.35** View of managers' workstation at Illinois facility (Author)

This facility also houses management and administrative functions for the region and could be regarded as "Class A" office space. The Iowa call center houses about 1000 employees, primarily call center agents. The facility is a renovated building that used to house "big box" retail center. It is an inexpensive, functional space that contrasts sharply with the high quality look and finishes of the Illinois location. Within this Iowa center, the agents are seated within pole and shell workstations clustered together in small groups and ordered in rows (see Figure 5.36).

**Figure 5.36** Cluster of agents' workstations at Iowa facility (Author)

Managers are seated in larger more fully enclosed workstations against the perimeter walls of the facility. It should be noted that this Iowa location has a higher density (about 10 percent greater) of workstations and people per square foot than the Illinois location.

### D. Participant Demographics

A total of 1900 surveys were made available through a secure Internet-based survey system to employees at both locations. A total of 600 surveys were returned for a return rate of 32 percent. Of the surveys returned, 301 came from the Illinois center and 260 came from the Iowa center. There were some differences in the characteristics of survey respondents at the two locations. However, from the standpoint of job functions within these groups, the work is essentially identical in nature at both locations, and the distributions of job types for respondents at the two locations were nearly identical. Regardless, these differences between the samples were statistically controlled in our analysis of the results of the survey.

## E. Types of Data Collected

For this study we collected a comprehensive range of subjective survey measures, and objective business and human resource metrics -- all data other than the survey data was independently collected as part of the organizations' ongoing business practices. The data collected included:

1.  Workspace/Behavioral Assessment (Survey)
2.  Automatic Call Distributor (ACD) Performance (
3.  Customer Satisfaction Scores
4.  Job Satisfaction
5.  Claims Data
6.  Lost Work Days

## F. Workspace/Behavioral Assessment (Survey)

The survey that was distributed to employees consisted of several sections, each of which contained questions pertaining to a particular dimension of workplace design, work behavior, or perceptions. We used a statistical analysis procedure called Factor Analysis to create the final groupings of questions. In this way, these groupings of factors emerged from the data, rather than being arbitrarily created by the research team. We then created indexes from each of these factors. Each index is a single summary number that is the mean of all the responses to the questions within the index. These indexes were then used in the means comparisons and other statistics that we calculated as part of the analyses. In all cases, with one exception, the individual questions making up the indexes, and the indexes themselves, used 5-point Likert-type scales in which a higher score indicates a more positive evaluation of that index. The one exception was the Discomfort scale, which used a 4-point Likert-type scale in which a higher score indicates a greater degree of discomfort. The eight indexes are as follows:

1. Workstation Features. This index consists of 15 questions that primarily address the fit of workstation design features to employee needs. The individual questions assess satisfaction with: arrangement of furnishings, technology tools, work surface size, lighting, noise, storage, privacy, confidentiality of materials, and availability of support services. A high rating on this scale indicates a high quality workspace.

2. Adjustability/Ergonomics: This index consists of six questions addressing the ergonomic design of the workspace. The individual questions assess

ability to adjust: storage, seating, shelves, lighting, keyboard, and monitor. A high rating on this scale indicates good ergonomic design.

3. Discomfort: This scale consists of six questions regarding pain or discomfort levels experienced on the job, related to comfort in back, shoulders, arms, hands, eyes, and legs. A high rating on this scale indicates a greater level of work-related discomfort or pain experienced on the job. Thus a low score on this index is more desirable.

4. Workspace Support for Collaboration: This index consists of six questions addressing the conduciveness of the workspace to collaboration with others. Questions included: availability of meeting spaces; spaces have required layout, furnishings, and technology; space permits quick shift from individual to collaborative work; and effectiveness of small meeting rooms. A high rating on this scale indicates that the overall workspace effectively supports collaboration.

5. Job/Environment Control: This index consists of three questions addressing the level of control over features of the workspace. The questions addressed ability to move/adjust features of the workstation, control over rate/speed of work, and control over job task decisions. A high rating on this scale indicates that the respondent has good control over the workspace.

6. Group Cohesion: This index consisted of four questions addressing cohesiveness within a workgroup. The questions addressed: degree to which group members are involved in group activities; access to required technology tools; and knowledge of how to use tools. A high rating on this scale indicates that cohesiveness within a workgroup is high.

7. Communication: This index consisted of four questions addressing ease of access and communication with co-workers. These questions included: quality of face-to-face communication, ease of accessing co-workers and supervisor, and ease of access to meeting spaces. A high rating on this scale indicates that employees have effective access to co-workers.

8. Community: This index consists of four questions addressing the professional image conveyed by the work space. The individual questions included: feeling part of the group; space conveys appropriate image to employees; workspace helps retain talented people; and workspace design makes people feel valued. A high rating on this scale indicates a high professional image for the workspace.

## 1. Comparison of Workspace/Behavioral Indexes between Call Centers

The results of the comparison survey scale analysis are shown in Table 5.43. The table shows the average rating on each scale obtained from respon-

dents at the two call center locations. Average index ratings that are significantly different at the two locations are highlighted in grey. The average
ratings are considered significantly different if the probability of observing
that difference by chance is less than .01 ($p < .01$). By this criterion, the
there is a significant difference in the average ratings at the two locations
on five of the eight survey scales (see Table 5.43). With the exception of the
Pain/Discomfort index, which has a range of 1 to 4 (4 being most pain), the
average ratings for each index have a range of 1 to 5 (with 5 being the most
positive rating possible).

There was no significant different on ratings of the Workstation Features

| Table 5.43. Significance tests for mean scores on workspace/behavioral indexes between locations | | | |
|---|---|---|---|
| **Index** | | | **ANOVA Calculation Significance Result** |
| | **Illinois Mean Score** | **Iowa Mean Score** | |
| Workstation Features | 3.24 | 3.16 | $F(1,535) = 2.2$   $p<.14$ |
| **Adjustability/ Ergonomics** | 2.17 | **2.24** | $F(1,519) = 7.69$   $p<.005$ |
| **Pain/Discomfort** | 2.31 | **1.91** | $F(1,511) = 24.22$  $p<.00001$ |
| Workspace Support for Collaboration | 3.62 | 3.59 | $F(1,496) = 0.29$   $p<.58$ |
| **Job/Environmental Control** | 3.39 | **3.85** | $F(1,473) = 35.84$  $p<.00001$ |
| Group Cohesion | 3.59 | 3.51 | $F(1,465) = 1.27$   $p<.26$ |
| **Communication** | 3.86 | **4.15** | $F(1,457) = 15.14$  $p<.0001$ |
| **Community** | **3.53** | 3.30 | $F(1,437) = 9.46$   $p<.002$ |

between the two sites (see Table 5.43). Thus, employee perceptions of satisfaction with: arrangement of furnishings, technology tools, work surface
size, lighting, noise, storage, privacy, confidentiality of materials, and availability of support services is equivalent at both locations.

The average ratings of the Adjustability/Ergonomics of the workstation
(ability to adjust: storage, seating, shelves, lighting, keyboard, and monitor)
at each site were somewhat unfavorable at both locations, but significantly
different between the two sites.

The Adjustability/Ergonomics rating of workstations at Iowa was significantly better than that of workstations at Illinois. Employee reports of work
related pain and discomfort (in back, shoulders, arms, hands, eyes, and legs)

was also significantly lower at Iowa than within Illinois. These results of better ergonomics ratings and lower pain at Iowa make logical sense; however, both locations could use improvement, given the relatively low ratings on this index. There was no difference between the two locations on ratings of Workspace Support for Collaboration (see Table 5.43). This result shows that both call centers appear to provide about an equivalent amount of meeting spaces; the meeting spaces have the same qualities of layout, furnishings, and technology; workspaces at both locations permit a quick shift from individual to collaborative work; and both locations have equally effective small meeting rooms.

An average rating of Job/Environmental Control (ability to move/adjust features of the workstation, control over rate/speed of work, and control over job task decisions) was significantly higher at Iowa than at the Illinois center (see Table 5.43). Better job/environmental control at Iowa may also be related to the significantly higher ratings of adjustability/ergonomics at that same location.

There was not a significant difference in average ratings of Group Cohesion (degree to which group members are involved in group activities; access to required technology tools; and knowledge of how to use tools) at the two locations (see Table 5.43). This index is primarily an assessment of group behaviors and technology. Group support behaviors are driven by management practices and enabling technology. Management practices, culture, and technology are probably similar at both locations.

Communication (quality of face-to-face communication, ease of accessing co-workers and supervisor, and ease of access to meeting spaces) is rated significantly higher at Iowa than it is at the Illinois center (see Table 5.43). Differences in communication are perhaps being facilitated by the openness and more organic feel of the workspaces and layout of space at the Iowa center. Differences in communication could also be influenced by supervisor management style although we assume that overall management style is similar at the two locations.

We found that ratings of sense of Community projected by the overall design (feeling part of the group; space conveys appropriate image to employees; workspace helps retain talented people; and workspace design makes people feel valued) is significantly higher at Illinois than at the Iowa center (see Table 5.43). The Illinois location is by any measure a significantly higher quality building in terms of design, layout, fit, and finishes used throughout. Our subjective perception is that Illinois feels like a local corporate headquarters (which in fact it is) and the Iowa location feels like a plain white box (it was a converted retail box store).

Other than the enhanced Sense of Community at the Illinois center, results of this comparison of survey indexes generally favors working conditions at

Iowa over those at the Illinois center. Iowa was rated as having significantly better Adjustability/Ergonomics, less Pain, better Job/Environmental Control, and better Communication than the Illinois center.

## G. ACD Performance Measures (Automatic Call Distributor)

We were able to obtain ACD measures of job performance for individuals at both sites. Three ACD performance measures were obtained:

- Average Handle Time (AHT): the average time (in seconds) it takes for the agent to handle a call. Generally, the lower the AHT score the better.
- After Call Work (ACW): the time (in seconds) for non-phone wrap up work resulting from a completed call. The lower the ACW score the better.
- First Call Resolution: the percentage of calls that are completed without having to be transferred to another resource for complete resolution. The higher the First Call Resolution the better.

These are commonly found measures used at most, if not all, call centers. Table 5.44 shows average AHT, ACW, and First Call Resolution measures for both centers for the 1 month period in which the survey was administered. Means for significant differences between locations are bolded. In this case, all differences in performance scores across sites are significant.

| Table 5.44. Comparison of average ACD performance measures at the two locations | | | |
|---|---|---|---|
| **ACD Performance Measure** | **Location** | | **Statistical Test Result** |
| | **Illinois** **N = 205** | **Iowa** **N = 198** | |
| AHT (seconds) | **398.6** | 481.3 | F (1, 401) = 23.79 p<.00001 |
| ACW (seconds) | 94.5 | **58.4** | F (1, 401) = 19.87 p<.00001 |
| First Call Resolution (%) | 89.1% | **92.5%** | F (1, 401) = 44.5 p<.00001 |

Illinois performed significantly better than the Iowa center on the measure of AHT while Iowa returned significantly better scores than Illinois on measures of ACW and percentage of First Call Resolution. These results may be highly dependent on the extent to which agents can actually control these performance measures. Agents probably have least control over AHT, since this measure is dependent on the customer's needs, and the difficulty of the calls received. Most managers agree that agents have more control over ACW and First Call Resolution. ACW can be lowered if the agent works in a workspace that allows efficient handling of post call wrap up, for instance, due to enhanced communication or control over the job (as is true for employees at the Iowa location). First Call Resolution can also be improved if the agent has the resources available in the workspace that makes it unnecessary to refer calls to a supervisor.

## 1. Workspace Predictors of ACD Performance Measures

We next performed analyses to determine how workspace characteristics across both locations affect the ACD performance of individual agents. The regression analysis involved examining the relationship between agents' ratings of their workspace (using the eight indexes shown in Table 5.43 as dependent variables, or predictors in the regression equation) and the ACD performance measures as the outcome measures.

## 2. Workspace and Behavioral Predictors of AHT (Average Handle Time)

None of the eight workspace/behavioral indexes was a significant predictor of AHT (Average Handle Time).

## 3. Workspace and Behavioral Predictors of ACW (After Call Work Time)

We found that three indexes: Pain/Discomfort, Group Cohesion and Community are significant predictors of ACW. The R2 value for the regression equation is .06, indicating that these three indexes predict 6 percent of the variance in ACW scores, a small but significant amount. Figure 5.37 shows an illustration of the relationship uncovered by the regression equation. In this figure, the numbers next to each arrow are the beta weights, which can vary between zero and .99, and show the relative strength of the relationship between multiple predictor variables. Figure 5.37 shows that pain and discomfort is positively related to ACW, thus as the amount of pain reported increases, so does ACW time. Agents in pain appear to take more time to complete a given amount of work than will agents who are not in pain. Group

Cohesion is negatively related to ACW. This is a desirable finding, because as Group Cohesion increases, ACW time decreases. Thus, agents who can rely on help from their peers can do the off-line handling of calls more efficiently (quickly).

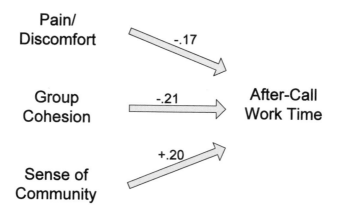

**Figure 5.37** Workspace and behavioral predictors of ACW (Author)

However, we also found that the sense of Community conveyed by the work environment has a strong positive effect on ACW. This result suggests that sense of community is probably not actually a detriment to ACW performance as much as it is simply not a useful or relevant predictor of this type of agent performance. Later in this report we reveal the important role that sense of Community plays in the findings.

4. Workspace and Behavioral Predictors of First Call Resolution

We found that one of the workspace/behavioral indexes, Workspace Support for Collaboration, was significantly related to First Call Resolution (percent of calls successfully completed without being transferred to other resources). The R2 value for the regression equation is .05, indicating that this index predicts 5 percent of the variance in ACW scores, a small but significant amount. Figure 5.38 shows an illustration of the relationship uncovered by the regression equation.

Thus, as Support for Collaboration increases, the percentage of calls transferred drops. Support for Collaboration is measured as a function of; availability of meeting spaces; meeting spaces have required layout, furnishings, and technology; space permits quick shift from individual to collaborative work; and effectiveness of small meeting rooms. This suggests that there is less need to transfer a call to other resources for resolution if there are optimal spaces and technology available to support collaboration in the call center. There was no significant difference in workspace Support for Collaboration between the two locations. However, this is a very important design concept that should be examined carefully in terms of Real Estate and workspace design strategy since it is a critical driver of costs to the organization and customer satisfaction.

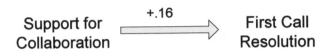

**Figure 5.38** Workspace and behavioral predictors of First Call Resolution

5. Comparison of Customer Satisfaction Scores between Call Centers

Customer Satisfaction scores are agent performance metrics that represent the average performance of a monthly sample of agents at a call center on the satisfaction of customers they have served. These scores give an objective indication of customer satisfaction with the quality of services provided by agents. This information is collected by an outside company from customers of the Telecommunications Company, by use of an automated survey. Thus not all agents have this data, because surveys are given to a random sample of agents within a given time period. Although we were not able to match these scores to the workspace/behavioral indexes (survey data) from individual agents (as we were able to do with the other analyses), we were still able to use this data to compare overall agent performance levels at the two locations and understand workspace design predictors.

We obtained three types of Customer Satisfaction scores from each location. These scores included percentage of calls resolved on the first contact (First Call Resolution), an assessment of agent performance during the interaction (knowledge, responsibility, efficiency, etc.), and an Overall rating of agent performance

For all three scores, a higher value indicates better customer satisfaction with agent performance. We analyzed 9 months of Customer Satisfaction results for the period preceding, and during, the time we were collecting survey and other data. Our analysis showed that each of the three individual satisfaction scores were higher for Iowa than for Illinois (means comparisons for: First Call Resolution (t = -2.15, p < .04) Entire Rep. (t = -2.97, p < .009) and Overall agent performance (t = -1.97, p < .06) are all statistically significant.

## 6. Workspace and Behavioral Predictors of Customer Satisfaction (CSTS) Scores

We calculated a regression analysis to determine if any of the workspace and behavioral survey indexes are predictors of average Customer Satisfaction scores. The results are shown in Figure 5.39. The results show that four indexes: Adjustability/Ergonomic features, Discomfort/pain, Job/Environmental Control and Community are significantly related to Customer Satisfaction scores. These four indexes predict 16 percent of the variance in those scores.The regression model shows that two workspace indexes; Adjustability/Ergonomics and Environmental/Job Control were positively and significantly related to the Customer Satisfaction scores at the two sites. The model also shows that as work-related pain and discomfort increase, these scores decrease. Last, as perceptions of image and community communicated by the workspace increase, Customer Satisfaction scores decrease. As in the ACW regression model, the image/community index has a negative, or non-helpful, relationship with the outcome measure.

## 7. Analysis of Job Satisfaction between Call Centers

We assessed employee job satisfaction at both locations using two commonly used questions, the first dealing with intention to stay, and the second recommend job to a friend. Overall, for both of these facets, Job Satisfaction is quite high at both locations.

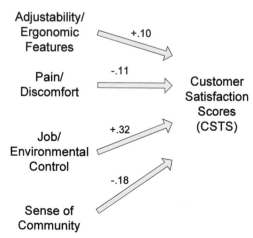

**Figure 5.39** Predictors of Customer Satisfaction Scores (Author)

***Intention to Stay.*** We found no significant difference in Intention to Stay responses between the two locations. However, Illinois employees are more likely to recommend a job at the Telecommunications Company to a friend.

8. Workspace and Behavioral Predictors of Job Satisfaction (Intention to Stay)

We computed a regression model, which shows that three indexes: Sense of Community, Job/Environmental Control, and Workstation Features significantly influence Job Satisfaction (Intention to Stay) (see Figure 5.40). The R2 value for the regression is .09, which shows that this model predicts 9 percent of the variance in Job Satisfaction (Intention to Stay) scores.

***Claims Data.*** We analyzed claims paid for injuries that occurred in the workplace at the Illinois and Iowa locations. There were a total of 72 accepted claims at the two locations, 57 for employees at Illinois and 15 for employees at Iowa. Illinios had 57 claims in a 36-month period, or about 1.5 claims/month. Iowa had 15 claims in a 19 months period, or about .8 claims/month. Thus the claims rate in Iowa is about half the rate in Illinios.

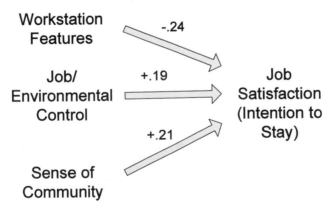

**Figure 5.40** Predictors of Job Satisfaction (Intention to Stay) (Author)

We computed ANOVAs to determine if there were differences in claims costs by job type and by location. Although there are large variations in average amount of the claims over job types, these differences are not statistically significant ($F_{(5, 63)} = -1.36$, $p < .24$). The large difference in average claims across the two locations is statistically significant ($F_{(1, 63)} = -3.59$, $p < .06$) and it certainly may be considered financially significant as well.

Table 5.45 shows the average amount of the settlement for all accepted claims broken out by location and the job type of the person making the claim. Although there are large variations in average amount of the claims over job types, these differences are not statistically significant ($F(5,63) = -1.36$, p<.24). Even the large difference in average claims across the two locations approaches but is not statistically significant by conventional criteria ($F(1,63) = -3.59$, p<.06) although it certainly may be considered financially significant.

| Table 5.45 Average claim amounts by job type and location (numbers in parentheses are the number of claims on which average is based). | | | |
|---|---|---|---|
| **Job Type** | **Location** | | **Job type Average** |
| | **Illinios** | **Iowa** | |
| Analyst | $425.16 (7) | $226.64 (3) | $365.60 |
| Associate Director | $121.00  (1) | - (0) | $121.00 |
| Coordinator | $7062.33 (4) | $6.00 (1) | $5651.06 |
| Customer Service Representative | $1074.85 (36) | $478.96 (11) | $935.39 |
| Sr. Customer Service Representative | $385.76 (8) | - (0) | $385.76 |
| **Overall Average** | **$1282.93** | **$396.96** | **$1098.36** |

Table 5.46 shows the average number or workdays lost, again broken out by location and the job type of the person making the claim. Although there are, as with the claims data, large variations in the average number of workdays lost over job types, these differences are not statistically significant ($F(5, 63) = -1.401$, p < .41). However, the difference in average workdays lost at the two locations is statistically significant ($F(1, 63) = -3.36$, p < .05).

## H. Interpretation

The results of this study suggest that the work environment at Iowa is superior to Illinios in terms of most of the dimensions of workspace quality measured in the survey. The work environment at Iowa is rated higher than Illinios in terms of Adjustability/Ergonomic design, degree that employees experience work-related pain, Control over the workspace and job, and Access to colleagues. The workspace at Illinios is rated higher than the one at Iowa only in terms of the degree to which the workspace design conveys

| Table 5.46 Average workdays lost by job type and location (numbers in parentheses are the number of claims on which average is based). | Location | | Job type Average |
|---|---|---|---|
| **Job type** | **Illinios** | **Iowa** | |
| Analyst | 1.85 (7) | 0 (3) | 1.3 |
| Associate Director | 0 (1) | - (0) | 0 |
| Coordinator | 16.25 (4) | 0 (1) | 13 |
| Customer Service Representative | 1.88 (36) | 1.54 (11) | 1.80 |
| Sr. Customer Service Representative | 2 (8) | - (0) | 2 |
| **Overall Average** | **2.84** | **1.13** | **2.48** |

a sense of community. In other works, the workspace at Illinios looks more businesslike but the workspace at Iowa is rated as better for doing business.

The differences between the Iowa and Illinios workspaces show up in terms of some measures of overall performance at the two sites. For example, the Iowa location had better performance than Illinios on average "after call work time" (ACW) and the percentage of calls that are answered without being transferred (First Call Resolution).

Evidence that the observed difference in performance at the two sites is a result of the difference in the quality of the work environment, rather than other differences between the sites, comes from our analysis of the relationship between individual performance measures and ratings by the same individuals of the quality of their workspace.

The results of the analysis show a significant relationship between four workspace quality ratings (Ergonomics, Pain, Cohesion, and Image) and individual performance measures. Lower Pain ratings were associated with lower average after call work times (ACW). Higher ratings of Cohesion were also associated with lower ACW measures. Image conveyed by the workspace was defined as: space helps employees feel like part of the group; space conveys appropriate image to employees; workspace helps retain talented people; and workspace design makes people feel valued.

Higher Image ratings were actually associated with higher ACW values, suggesting that the quality of the interior space may not be influencing agent performance outcomes, although we did find later in this study that image has important relationships with job satisfaction, which in this case is a leading indicator of retention.

The advantage of Iowa over Illinios in terms of ratings of workspace design shows up most dramatically in terms of independent measures of customer satisfaction scores. Over a 9 month period the average Customer Satisfaction scores for Iowa were significantly higher than those for Illinios. Iowa also did better than Illinios in terms of workspace injury claims. The average costs of claims paid as well as the average number of workdays lost due to injuries were lower at Iowa than at Illinios. Although these differences were not statistically significant (largely due to the small number of cases) they are consistent with the other findings of this research, showing the superiority of the workspace at Iowa over that at Illinios in terms of worker satisfaction and productivity.

## IX. SOME THOUGHTS ON PRODUCTIVITY

These Case Studies represent another step towards relating some basic measures of productivity with features of the work environment. The purpose in pursuing a program of research to examine productivity is twofold: First, to show that the physical work environment can be thought of as a strategic business tool (much as the investment in human capital and IT are) and, second, to provide credible evidence that broad workplace strategies to support behaviors (such as collaboration) that managers intuitively believe lead to effective work, actually do increase performance and business success.

Productivity is defined, at its most basic, through measures that relate quantity and quality of output to input of labor. Labor may be physical or intellectual. Productivity and profitability are linked in that, all things being equal, a company that can raise its productivity will be able to lower its costs and raise its margins. However, productivity by itself rarely ensures business success and should not be the sole driver of investment in the physical office space. Providing value to the customer is often more important than increasing productivity. Thus, we believe that metrics that go beyond process time and cost but contain measures of quality of the output of the process, especially *from the customer's perspective,* are critical to providing a balanced view of productivity.

The results of this study indicate that the investment made by the organizations described in this Chapter has resulted in significant improvements in individual and group work effectiveness, and measurable enhancements in efficiency of business processes. The Case Studies described in this chapter represent design strategies that support the appropriate departmental and group adjacencies, meeting spaces to support collaboration, and the overall quality of work products. At the same time, the approaches outlined in the Case Studies support a clear Real Estate strategy of cost reductions through reductions in owned or leased square footage. Despite the cost reductions, these studies show that investment in appropriate furnishings and interior design of space can continue to support and enhance work effectiveness.

.

# CHAPTER 6

## Case Studies: Collaborative Spaces and Team Performance

## I. INTRODUCTION

A continuing trend in the workplace is the use of groups and teams in various forms to accomplish higher quality work in less time (Jones et al., 1993; Becker, 2004; Davenport, 2005). Although various designers and manufacturers have attempted to design collaborative work spaces to support this new organizational structure, there is little published quantitative research that compares the effectiveness of these new space forms on supporting this work.

This chapter begins with background on the use of teams in business organizations, including a typology of teams. Next, we discuss a model of the effects of environmental control as it relates to group work. This is followed by five Case Studies selected from recent projects conducted by the author and research team. In each Case Study, we examine the effects of environmental control, implemented in various ways through flexibility of design of workstations and adjacent group spaces.

*Case Study 1.* In the first Case Study, we investigated the influence of highly adjustable (and moveable) workstations on the effectiveness of three different project engineering teams (a total of 45 people) within a high-technology manufacturing company. The key measures for this study included: job and environmental control, individual performance, group effectiveness, communication and collaboration, stress and musculoskeletal pain.

*Case Study 2.* In the second Case Study we examined the effects of team work spaces on sales team effectiveness (a total of 30 people) in three different sales offices. One site served as the "experimental" group, in which changes were made, and the other two locations served as a control group. The key measures for this study included: individual performance, group effectiveness, group participation, stress, adjustability of workspace, work demands, job satisfaction, job control, and musculoskeletal problems.

*Case Study 3.* In the third Case Study, we examined the effects of more open workstations and related collaboration spaces on the effectiveness of 90 research and development employees (an additional 90 people served in a Control group). The key measures for this study included: collaboration, internal group process, sense of belonging, and identification with the organization.

**Case Study 4.** In the fourth Case Study, we examined the effects of work team spaces on the effectiveness of Claims Processing teams within an insurance company located in the Midwest. This is highly repetitive work that is usually associated with high workload demands and stress. We redesigned the spaces for one team of four people. The key measures for this study included: time spent in workspace, workspace efficiency, individual performance, group effectiveness, stress, musculoskeletal complaints, visual problems, control over workspace, control over pace of work, privacy, and stretching and twisting.

**Case Study 5.** In the fifth Case Study, we examined the impact of group work spaces on collaboration and sense of community for professional workers within an organization in the consumer foods industry. A total of 88 employees (half of which served in a "control" group of people who did not receive new spaces) participated. The key measures for this study included: a workspace evaluation, collaboration, group work process, communication, retention, and sense of community.

## II. TEAMS AND TEAMWORK

Organizations have clearly expanded their use of teams and teamwork to accomplish work. The use of teams as a building is fundamental to the way work is organized and managed (Davenport, 2005; Gordon, 1992; Cohen, 1993; Bader, Chang and Bloom, 1999).

### A.      Forces Leading to the Use of Teams

Organizations create and deploy teams at all levels, to meet tactical and strategic goals. Although teams of various forms have been in existence for many years, what is more recent is the integration of teams into the formal organizational structure, and their level of accountability for business unit goals. The basic assumption behind the use of teams is to leverage the synergy that exists in any group of people -- to create a gestalt in which the whole is greater than the sum of its parts. A variety of "change drivers" have compelled organizations to implement teams, including competition and new technologies that support teamwork.

### 1. Competition

The forces of global and regional competition have resulted in the rapid compression of product development cycles (both physical products and service products) and have forced companies to focus considerable attention on issues such as product quality and efficiency of distribution channels.

To support these trends, considerable effort has been placed on driving authority and autonomy "down" into the organization, so that people close to the problem can take ownership and action to meet their objectives. When such responsibility is reallocated, it typically resides in teams. Cross functional product development teams are one example of organizational response to competition. Such cross functional teams have been shown to greatly reduce product development cycle time in conjunction with manufacturing and marketing efforts. Teams have also formed around quality improvement issues. Self-managed teams have also appeared, particularly in manufacturing settings, in which groups of workers make decisions previously made by group managers.

## 2. Information Technology

Information technology continues to play a central role as a change driver in business. It both creates the opportunity for, and facilitates, teamwork. A variety of groupware software products, running on LAN and Intranet sites, are now commonly employed by organizations to enable teams of workers (physically dispersed or otherwise) to effectively communicate and move projects forward. Obviously, the ability to manipulate data pertaining to the business and its mission facilitates the quality of team decision-making. Technology allows scarce and expensive specialists to efficiently contribute to teams on an as-needed basis. The great increase in business travel, which can tend to slow the work progress of even co-located team members, can be supplemented through appropriately designed information technology. The speed of decision making of teams is greatly accelerated through the use of technology. As long as the quality of team work is maintained, the result will be a competitive advantage to the organization.

## 3. The Future of Teams

Team organization will move in two countervailing directions (Cohen, 1993).

Relatively permanent, self-managed teams will continue to be organized around products, services, and customers. These teams are the units that "get the work done" in organizations.

Companies will also move outside their traditional boundaries by employing networked structures in which team members assemble to handle temporary assignments, and then disband when the work is complete. These looser, collaborative structures are fluid and responsive to change.

In general, organizations have seen the competitive advantage of structuring using teams and despite the obstacles (such as performance measurement and compensation) that can hinder implementation of teams, continue to employ this tactic. In the following section, we discuss four different team types now commonly found in organizations.

## 4. Types of Teams

In this section, we discuss four types of team designs that are employed by organizations. These team designs include: Networked (also referred to as "fishnet" designs, see Johansen and Swigert, 1994); Parallel teams, Project team designs, and Work teams (Cohen, 1993). Networked teams are composed of individuals or entire teams that are loosely networked together for some period of time in order to accomplish some goal. Parallel teams are made up of people who monitor the performance of processes or products and make recommendations for improvements in parallel to actual work. Project teams have the responsibility for completing projects or assignments that typically last some extended period of time. Work teams are typically designed as a group of people who have overall responsibility for some function or product.

These designs can be arrayed on a continuum from informal and temporary to relatively more formal and permanent.

*Network design.* This design is made up of the interrelationships between individuals or groups of individuals who cooperate to achieve a common purpose. The "nodes" of the network are occupied by individuals or groups, and the links between the nodes represent interactions between the individuals or groups. Networks may be tight or loose, depending on the intensity of interaction between the nodes. Not all the nodes within a network are linked, and some nodes have multiple links (O'Neill, 1991b). The purpose or business goal of a network is its reason for existing. There is always some problem to be solved or objective driving the network.

Networks are different from other team structures in that the boundaries between the network and the organization are not clearly defined. Membership in networks is fluid and diffuse. Members of the network may only be able to identify the other members with which they have direct contact. The network itself extends beyond individual team boundaries.

Examples of networked designs are found in professional service firms such as consulting organizations and, to some extent, in high technology companies. Any organization that must perform highly complex tasks in the context of shifting market demands and customer needs may be organized as a network. Several authors have suggested that networked designs are the wave of the future (Johansen and Swigert, 1994).

*Parallel Team Designs.* Parallel team structures supplement normal work activities and are usually temporary (Cohen, 1993). They carry out functions that the existing teams are not equipped to carry out well (Lawler, Mohrman, and Ledford Jr., 1995). Examples of parallel teams include quality circles, productivity improvement groups, tasks forces, and the like. Parallel teams have clear boundaries and can identify team members from non-members. These teams typically make recommendations for change that are then considered by the management of their organization. Organizations tend to implement parallel teams fairly often, but only a small number of employees are actually involved in parallel teams at a given time (Cohen, 1993). Parallel teams have trouble achieving legitimacy within the organization because they are short lived and have little political power or control over a budget. They usually compete with the organization for time, money, and other resources. There is little empirical evidence regarding the effectiveness of this type of team.

***Task Forces.*** Task forces are another common type of parallel team, but they differ from quality circles in some significant ways. First, they are usually asked to make recommendations to solve a particular problem, as opposed to generating general ideas for improving quality. Second, membership may be mandated rather than voluntary. Third, they usually have a specific deadline for accomplishing their mission. Fourth, task forces are used at all levels within the hierarchy of the organization. Task forces are common because they represent an effective way of getting people to think "outside the box" and generate novel ways of solving problems. They are flexible and responsive to change (Cohen, 1993).

***Project and Development Teams.*** This type of team is common within organizations and is typically made up of professional people such as designers, engineers, researchers, marketing people, and others to fulfill users' needs within an extended but defined time frame (Cohen, 1993). Examples of project teams include: new product development teams, information system teams, R&D teams, and others.

Project teams are assigned unique tasks for which the exact outcome is unknown. This type of team usually has a mandate from management, along with a budget and the political power to get things done. The teams themselves are usually self-managing, but must respond to the requirements of their sponsor and ultimate customers for their product. The purpose of the team is usually aligned with the strategic objectives of the business.

Project teams are different from the task force in several ways. First, their work is integrated into the main purpose of the organization (for instance, to make products). Second, they tend to have the authority to make decisions and the resources to carry them out. Third, the life span of project teams is typically longer than that of task force teams, although with shortening product development cycles, this temporal differentiation may be changing.

***Work Teams.*** Work teams have the overall responsibility for producing products or services (Cohen, 1993). In contrast to parallel or project teams, they perform regular, ongoing work. Work teams may perform their own internal functions, such as hiring and firing decisions, determining pay changes, etc. The people in a team may work together in terms of moving the mission of the group forward, as well as have individual responsibilities for projects or other tasks that are related to the mission of the Work team. Ultimately, the members of a work team have shared performance. Work teams can be self-managed (which is increasingly typical of teams in manufacturing settings) or have a manager.

## B. Conclusion

This section has briefly addressed the concept of different team designs and provided examples of four general types of teams. This discussion is not meant to be exhaustive, and there are many books the reader can refer to for extended discussions of teams and their implementation. However, it is clear that teamwork in various forms is here to stay in the U.S. and around the world, and that the competitive and technological drivers of teamwork in the U.S. continue to both support and push the use of teams within organizations.

The continuing theme of this book is the notion of worker control over the environment and the relationship between control and organizational effectiveness. In this chapter we continue that thread by discussing our model of environment control as it applies to team work, as well as some research and case studies that relate aspects of control to team and group settings.

## III. MODEL OF CONTROL FOR WORK TEAMS

The potential for environmental control at the work group or team level is determined by the flexibility of the components within the workstation, as well as the physical (and organizational) boundaries surrounding the team. Boundaries serve to control the flow of information to and from the group. Control may be exercised by the ability to determine and self manage the reconfiguration of work space layout and boundaries as group mission dictates. Organizational boundaries can be reinforced or made more permeable through the design of the physical work space. Boundaries can be used both to integrate the team into the larger organization, or differentiate the team from the organization. Research indicates that the way teams manage interaction across their organizational boundaries influences satisfaction and group effectiveness. When the team mission depends on external integration, effectiveness can depend on pace and timing of information exchanges with other units. Team effectiveness can also hinge on the ability to isolate certain activities from outside interference, such as problem solving meetings or sensitive R&D areas. The model suggests that both the integrative and differentiating functions can be supported by team control over the layout of the work environment.

Control over layout and boundaries may also influence inter-member communication and cohesion. Face-to-face interaction may be influenced by proximity of workstations and meeting places. Quality of communication has been linked to performance outcomes in teams. Communication and exchange of information is of prime concern for team members. Johansen and Swiggert (1994) provide a framework for understanding team member interaction and communication based on place and time (see Figure 6.01).

This framework suggests that teamwork can take place in four combinations: Same Place/Same Time, Same Place/Different Time, Different Place/Same Time, and Different Place/Different Time (Johansen et al., 1991) (see Figure 6.01). The Same Time/Same Place mode of group activity is a way of working that has been around for as long as people have been working in groups. Examples of this work style include face-to-face meetings that may be planned or informal. Many important meetings occur in informal, unplanned spaces, such as hallways and coffee bars. The Different Time/Different Place work style is also fairly common, except that communication is asynchronous rather than simultaneous, because the participants are at different locations. The media of communication is typically paper reports and memos, but voice mail and e-mail have also become common methods. The case studies and research that we discuss in this chapter reflect teamwork that takes place in the "Same Time, Same Place" mode, which is common to every organization.

## A. Job Control

One job characteristic widely thought to influence performance, stress, and health is the amount of control the worker has over the job. This concept is known as "job control." An important component to job control is "decision latitude," which is the degree of freedom the worker has to make decisions about how to do the job. The "demand/control" model, originally conceptualized by Robert Karasek, incorporates this notion of job control. The demand/control model suggests that "demand," which is the psychological demands of a particular job (how hard you have to work), coupled with the level of job control, can predict worker stress and health.

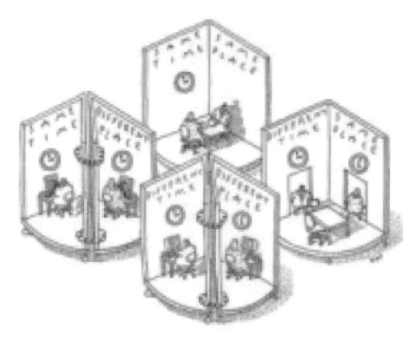

**Figure 6.01** Framework for location and time interaction of team members (permission Herman Miller, Inc., copyright 1993)

For example, the model predicts that fatigue, anxiety, and physical illness are more likely to occur when psychological demands of a job are high and job control is low (Karasek and Theorell, 1990).

A substantial amount of research has linked job control with factors such as performance (Glass, Singer and Pennebaker, 1977), stress (Landsbergis, 1988; Perrewe and Ganster, 1989), and health (Johnson, Stewart, Hall, Fredlund, and Theorell, 1996; Karasek, Gardell and Lindell, 1987; Semmer and Frese, 1988). Glass et al., (1977) found that performance of complex tasks was higher and error rate was lower when workers had control over an external stressor (an aversive noise). Other studies have shown that job control is linked to reduced stress and/or improved health (Karasek, 1979; Hedge, 1988; Uehata, 1991). The cumulative effects of psychological stress can be serious, leading to health problems such as coronary heart disease (CHD). Hedge (1988) found that perceived job stress among office workers was highly associated with work-related illnesses such as lethargy, headache, stuffy nose, and dry throat. Johnson et al. (1996) demonstrated, using a sample of 12,517 Swedish males, links between low job control and cardiovascular mortality that remained significant even after controlling for such influences as social class, education, smoking, exercise, or age.

## B. Environmental Control

Most studies to date have conceptualized job control as the ability to make decisions about such things as the pace of work or how work is accomplished. In the present study, the concept of control is applied to the physical environment. We believe that the provision of control over the work environment can have similar positive effects on worker health and effectiveness. Our model suggests that work environments possessing a high degree of flexibility and user-adjustable features will enhance the availability of environmental control. Increased opportunities for environmental control should allow the worker to modify the work environment in response to changing work flow, tasks, and job demands. Thus, the availability of environmental control (through work environment adjustability and mobility), coupled with knowledge of how to exercise control (through training on how to make adjustments) may support work flow and enhance worker effectiveness and health.

## IV. CASE STUDY: EFFECTS OF A HIGHLY ADJUSTABLE WORKSTATION ON PROJECT TEAM EFFECTIVENESS

### A. The Project

Many organizations are seeking ways to use the work environment as a means of improving worker effectiveness without sacrificing physical or psychological health. In addition, teamwork and group performance have become important issues for businesses that rely on collaborative efforts to meet organizational goals. These trends have put pressure on organizations to provide flexible work environments to meet the needs of project-based teams and other small groups whose size will vary over the course of a project. There appears to be an ongoing need for work environments that can support the goal of enhancing individual and group effectiveness without increasing risks to worker health.

In the present study, we investigated the influence of control over the physical environment on worker effectiveness, team effectiveness, and worker health. The "H-A" ("High-Adjustability") workstation was designed as a conceptual prototype with the intent of providing a great deal of physical control via a highly adjustable physical work environment (see Figure 6.02). The workstation interior is highly adjustable, having the ability to adjust the primary work surfaces from seated to standing height (see Figure 6.02, left and right images). Within the workstation, there are a number of highly adjustable work surfaces which themselves are height and angle adjustable (see Figure 6.03).

Design for environmental control was intended to enhance users' individual and group performance, and to reduce stress and musculoskeletal discomfort. The H-A was also designed to allow users to reposition or relocate their workstations. The intent of this mobility was to better support teamwork and collaboration between individuals as well as address facilities space-use issues.

This Case Study presents the results from observations of 40 workers, 20 who received the H-A workstation and 20 who served as a control group and received standard "cubicle" workstations. We hypothesized that the H-A would increase availability of control and workers' knowledge of control over their workstations. We also predicted that availability of control and knowledge of control would improve individual and group effectiveness, enhance group collaboration, and decrease stress and physical health problems. Survey and interview data were gathered before and after the installation of the experimental workstations to permit evaluation of the impact of design features on the outcome variables.

## B. The Study

### 1. Hypotheses

The present study investigates the effects of a work environment designed to be highly adjustable, and mobile, on outcomes including individual performance, group effectiveness, and health (as measured by psychological stress and musculoskeletal problems).

We predicted that use of a mobile, adjustable workstation design (availability of control), paired with ergonomic training (knowledge of control) would enhance health and effectiveness.

Figure 6.04 shows a conceptual model of the relationships between the variables we tested in this study. The study hypotheses, summarized here, are represented by the model (Figure 6.04).

*Hypothesis 1:* The provision of a flexible physical work environment (i.e., The H-A workstation) paired with ergonomic training will increase the availability of environmental control and user's knowledge of how to control the environment.

*Hypothesis 2:* There will be a positive relationship between control indices (availability of control and knowledge of control) and individual performance, group collaboration, and group effectiveness.

**Figure 6.02** Illustration of "H-A" workstation (Author)

*Hypothesis 3:* There will be a negative relationship between control indices (availability of control and knowledge of control) and job stress and musculoskeletal discomfort.

## 2. Methods

*Work Environment.* The study took place at a multinational technology manufacturing company located on the west coast of the U.S. The site was previously a warehouse facility that was refurbished so that all aspects of the manufacturing process (design, development, manufacturing, etc.) are housed in one building. Renovation of the building occurred just prior to the start of this study. This facility houses approximately 1200 residents. Of these occupants, 400 are manufacturing workers and 800 are office workers that were relocated from other sites. Those in the office environment had been using various brands of office systems furniture products.

*Participants.* Three different functional departments within the company were involved in the study, all related to various aspects of product development engineering. All of these workers were organized into Project Teams. These included Design, Cost, and Manufacturing Engineers. We included people with varied job types to see if the H-A workstation (and the enhanced level of control we hypothesized it provides) was a better solution for one job type than another. If the results were the same for all groups, we would then have some assessment of the generalizability of the results to a fairly broad range of job types.

Participants who were to receive the H-A workstation were randomly selected from the three work groups: 9 from the Design group, 12 from the Cost group, and 19 from the Manufacturing Engineering group.

The sample consisted of 27 men and 13 women between the ages of 23 and 61 (median = 36, SD = 6.99). Within each of the work groups, half of the participants were randomly assigned to the Experimental condition (H-A workstation). The other half served as a Control group and received standard systems furniture workstations. Overall, there were 21 participants in the Experimental condition and 19 in the Control group.

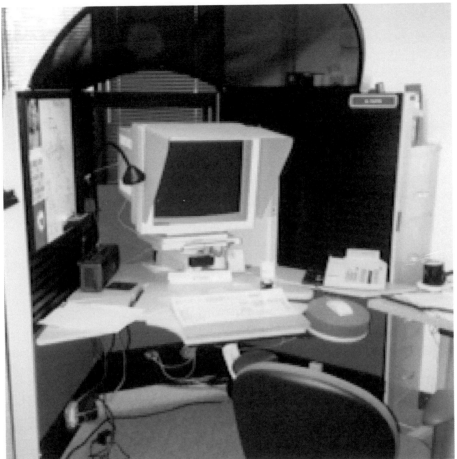

**Figure 6.3** Photo of "H-A" workstation work surfaces (Author)

## 3. Research Design

The research used an "untreated control group design with pretest and posttest" which is used frequently in field research (Campbell and Stanley, 1966). See Figure 6.05 for a graphic depiction of the study design, (O = observations and X = the introduction of the H-A workstation). The first observations were done on both groups simultaneously before any changes to the work environment were introduced ($O_1$). Next, the H-A workstations were installed for the Experimental group (X).

At that time, the Control groups received the new systems furniture workstations that are the corporate standard for that facility. Within each functional department, the Experimental and Control groups were located in the same physical area. Three months after the participants in the

Experimental groups were moved into the H-A workstations, both groups were again observed (see Figure 6.05, "O$_2$").

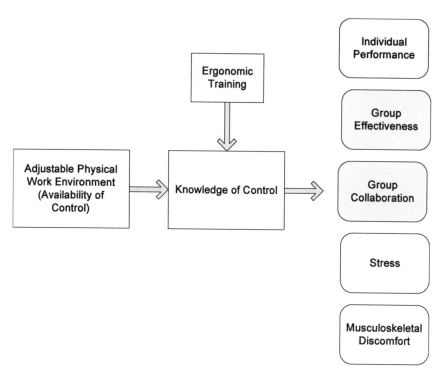

**Figure 6.04** Conceptual Model of Study Variables (Author)

| Experimental Group | O$_1$ | X | O$_2$ |
|---|---|---|---|
| Control Group | O$_1$ | X | O$_2$ |

**Figure 6.05** Untreated Control Group Design with Pretest and Posttest (Author)

This research design made it possible to ascertain whether changes in the test group were due to the H-A workstations, and rule out possible threats to the internal validity of the study. Two additional, informal observations took place at 3-month intervals after O$_2$ during which semi-structured interviews were conducted and workspaces were photographed.

## 4. Training

When participants received their new workstations, they underwent thorough training on how to adjust the workstations, their operation in terms of movement, and on the importance of ergonomics in the workplace. Part of the training involved a slide presentation that featured four major topics: design of individual workspace, layout of total work environment, job design, and the overall relationship between people, technology, and organization.

The training also included a demonstration of the features in each workstation and instructions on adjustability to a preferred configuration for the user. Finally, the trainer and worker collaborated to adjust the work area to fit the needs and anthropometrics of the user. Both H-A users and workers using standard systems furniture received similar training and adjustments (to the extent possible).

## 5. Data collection

A questionnaire was distributed prior to the installation of the H-A workstation, and again three months after the change. Portions of the survey were taken from Brill, Margulis and Konar (1984), Caplan, Cobb, French, Harrison and Pinneau (1975), Nadler (1977), and O'Neill (1994). The questions on the survey investigated a broad range of issues including: satisfaction with workspace design, lighting, privacy, measures of individual performance, group collaboration, group effectiveness, job demands, job stress, and questions measuring musculoskeletal discomfort.

We also developed an objective, 20-item checklist, developed to determine the number of adjustable furnishings within the workstation. This checklist was used to index the availability of environmental control within a workstation.

In addition to the questionnaires, employees were interviewed using a semi-structured survey protocol. They were interviewed both before implementation as a needs assessment and after implementation at three-month intervals as a means of collecting qualitative insights and comments regarding the overall work environment. This information was used to complement the quantitative data collected through the questionnaire.

## 6. Data analysis

A Principle Components Factor Analysis (PCA) with Varimax Rotation produced a 15-factor solution that accounted for 53.33 percent of the total variance. Nine factors were selected from the original 15 factors for further analysis. Five of these, including: storage, privacy, lighting, learning, and decision latitude, were included as predictor variables to be used along with the environmental control indices to predict the outcome variables. The other four factors, including: individual performance, group effectiveness, group collaboration, and stress, were designated as outcome variables.

The reliability of these constructs was further tested using Chronbach's alpha ($\alpha$), a statistic that allows researchers to ask a number of questions relating to the same concept (for example, performance) and then statistically collapse the responses to these questions into a single factor representing that construct. Chronbach's alpha can range from 0 (representing a completely unreliable index) to 1 (meaning a highly reliable and useful index). The factors developed for this study all had acceptable alpha levels.

## 7. Study Variables

***Predictor Variables Operationalized.*** The Predictor Variables are as follows:

Storage ($\alpha$ = .79) consisted of four questions that measured the extent to which respondents were satisfied with the storage provided by their current workspace.

Privacy ($\alpha$ = .73) consisted of four questions designed to measure both auditory and visual privacy.

Lighting ($\alpha$ = .75) consisted of two questions measuring satisfaction with lighting including workplace shadows and glare.

Learning ($\alpha$ =. 80) consisted of five questions measuring the degree to which one's job requires skill development and creativity.

Decision Latitude ($\alpha$ = .62) consisted of three questions assessing the freedom to make decisions about how to do one's job.

***Outcome Variables Operationalized.*** The Outcome Variables are as follows:

Individual Performance ($\alpha$ = .94) consisted of nine questions that asked the respondent to rate his/her performance for the last three months. Performance items related to issues such as amount and quality of work, meeting deadlines, and creativity.

Group Effectiveness ($\alpha$ = .95) consisted of seven questions that asked for ratings of one's work group on dimensions such as resolving conflict and making decisions.

Group Collaboration ($\alpha$ = .93) consisted of six questions that measured the extent to which one's work group was supportive, participated evenly, and shared responsibility.

Stress ($\alpha$ = .86) consisted of seven questions looking at how bothered, worried, frustrated, and tense one feels at work.

***Control.*** Based on the results from the checklist of furniture adjustability, two indices of control were created as follows:
Availability of Control
This variable was measured by the number of adjustable features in the workstation divided by the total number of features or accessories in the workstation.
Knowledge of Control
This variable was measured by the number of features the end user knows how to adjust divided by the number of adjustable features in the workstation.

***Musculoskeletal Measures.*** Finally, from the 27 musculoskeletal discomfort items in the survey, 5 discomfort factors were created:
- Head/Neck Discomfort
- Upper Limb Discomfort
- Leg Discomfort
- Hip Discomfort
- Torso Discomfort

The model was tested using Analysis of Variance (ANOVA) and multiple regression techniques. First, an ANOVA was conducted with the control indices to compare participants who received the H-A workstation to those who received the systems furniture workstations. Then, multiple regressions were used to learn whether control was a significant predictor of the outcome variables.

8. Results

***Hypothesis 1.*** In Hypothesis 1, we conjectured that the use of the H-A workstation paired with ergonomics training would increase availability of control and knowledge of control. Availability of control was defined as the ratio of adjustable furniture in one's workstation to the total number of furniture items in the workstation. Knowledge of control was defined as the ratio of workstation features that a user knows how to adjust to the total number of adjustable items in the workstation.

A two-way ANOVA, workstation type (H-A, Systems Furniture Workstation) by data collection point (before installation, 3 months post-installation) was calculated. The results of this analysis showed that our hypothesis was supported -- use of the H-A workstation, and ergonomics training, increased both availability of control and knowledge of control (see Figures 6.06 and 6.07).

We found a significant interaction on workstation type which illustrates that the availability of control increased greatly as a result of the installation of the H-A workstation (see Figure 6.06). Thus, we found that the move to the H-A caused an increase in the availability of control for the experimental groups (F (1, 71) = 40.4, p < .001, see H-A line, Figure 6.06). The availability of control for those who received systems furniture cubicles, however, remained stable over the course of the study (see Standard Cubicle line, Figure 6.06).

This is what we expected because workers in the control group were moved to new office systems furniture that was similar in design and function to the furniture they used before the start of this study.

There was also a significant interaction on workstation type for the "knowledge of control" factor.

As shown in Figure 6.07, for those who received the H-A workstation, there was a clear increase in knowledge of control during the study. For those who received systems furniture cubicles, however, a sight decrease in knowledge of control was noted.

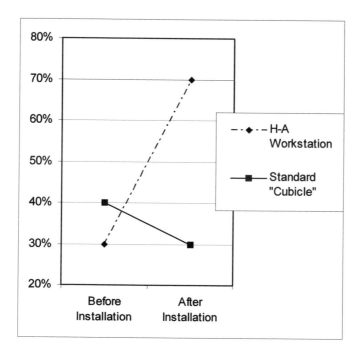

**Figure 6.06** Availability of control - Percent of all work features identified as adjustable (Author)

All workers had received training on how to modify the adjustable features in their workstations. Our findings indicate that those who received the H-A work environments were more confident in their knowledge of how to adjust the workstation.

This may be because the highly adjustable H-A workstation offered users more opportunity to apply and reinforce the knowledge they gained through training, and through interaction with the workstation.

*Hypothesis 2.* Our second hypothesis was that there would be a positive relationship between control (both availability and knowledge of), and the outcome variables of individual performance, group effectiveness, and group collaboration. We tested this hypothesis using a series of multiple regression calculations. In each case, we found a statistically significant relationship between availability of control or knowledge of control and the outcome variables (see Figures 6.08, 6.09, and 6.10). In each case, as aspects of control over the physical environment increased, individual performance, group effectiveness, and group collaboration increased.

In each of Figures 6.08, 6.09, and 6.10, the b values represent the relative contribution of each of the predictor variables to the variance in the outcome measure. Thus, in Figure 6.08, decision latitude and knowledge of control predict individual performance. The model shows that as decision latitude and knowledge of control increase, individual performance increases. The model shows that decision latitude b = .316) and knowledge of control (b = .322) contribute about equally to level of individual performance.

The models in Figures 6.09 and 6.10 both show that the availability of control significantly influences both group collaboration and group effectiveness. The model for group effectiveness is particularly interesting in that the predictor variables include control over both the physical and psychosocial work environments. Ability to learn on the job is also a predictor of group effectiveness.

*Hypothesis 3.* Multiple regressions were also used to assess the third hypothesis, which stated that there would be a significant negative relationship between the control indices and the outcome variables of stress and musculoskeletal discomfort. In other words, we predicted that greater control would be related to reduced levels of stress. As shown in Figure 6.11, this hypothesis was partially supported by a significant finding for the stress variable. This figure reveals that, as knowledge of control, privacy, and satisfaction with storage increase, psychological stress decreases. This finding was not replicated with any of the musculoskeletal variables, which may have been due to the low number of reports of musculoskeletal discomfort.

Overall, we found that knowledge of control (through training) and availability of control (through adjustability of the work environment) reduced stress and increased group collaboration and effectiveness for team workers.

## 9. Analysis by Work Group

By including three distinct work groups in this study, we hoped to learn whether the response to the H-A Workstation was equivalent across job types and work teams. We investigated this issue by using a workstation by work group ANOVA on the 3-month post-installation survey data. All predictor and outcome factors were tested as dependent variables. Overall, there were few significant findings, indicating that the principal findings of this study are generalizable across work groups. Two variables that did produce different results depending on work group, availability of control, and satisfaction with storage, are presented in Figures 6.12 and 6.13.

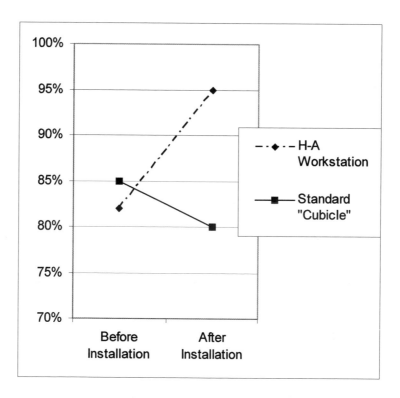

**Figure 6.07** Knowledge of control - Percentage of adjustable features users reported knowing how to adjust (Author)

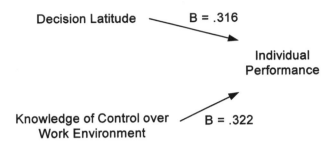

**Figure 6.08** Individual Performance, R2 = .31, p < .01 (Author)

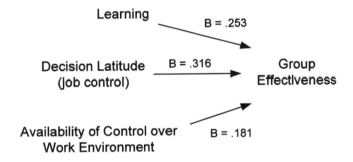

**Figure 6.09** Group Effectiveness, R2 = .242, p < .001 (Author)

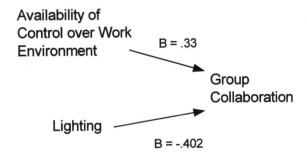

**Figure 6.10** Group Collaboration, R2 = .24, p < .001 (Author)

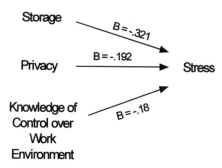

**Figure 6.11** Stress, R2 = .213, p < .01 (Author)

*Availability of Control.* Figure 6.12 shows that availability of control was always greater with the H-A work environment than with systems furniture cubicles (F (2, 32) = 5.24, p < .05), and this difference was most pronounced for the Manufacturing Engineers.

Interviews with the engineers illustrated their feelings about the adjustability of the H-A workstation. One noted, "It is great to be able to work in sitting and standing operations. My neck and back and wrists do not hurt as much. Also, I like the ability to change my workstation configuration easily and several times a day." Another engineer said, "I have moved my workstation several times and it seems to work really well."

One explanation for the observed difference in availability of control among work groups using the H-A workstation may be due to the different abilities of the groups. Perhaps manufacturing engineers understood that more features are adjustable due to their understanding of mechanics and design. Another possible explanation relates to the management styles present in the different work groups. The manager of the manufacturing engineering group was very open to the installation of the H-A and gave the work group a great deal of freedom to decide how to configure the work area. Managers in the other two work groups placed more constraints on the organization of the workstations within the work area.

*Satisfaction with Storage.* Overall, standard cubicles received higher ratings on satisfaction with storage (F (2, 31) = 12.56, p < .001). Figure 6.13 illustrates that, for the Design and Cost Management groups, standard cubicles received higher ratings on satisfaction with storage (on a 5-point scale, with 5 being the best rating).

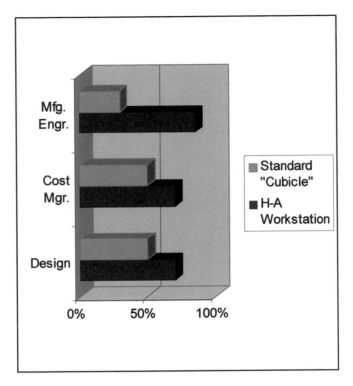

**Figure 6.12** Availability of Control (Author)

However, for the Manufacturing Engineers, the H-A was given higher ratings on satisfaction with storage. The comments from the Design and Cost Management groups indicated that they felt the H-A workstations did not provide enough storage space.

Several workers complained that the H-A workstation did not have enough shelves. In addition, one worker from the Physical Design group felt that file storage in the H-A was inadequate and said, "We need to have some filing system that folds under the workstation where you could file frequently used papers and references." The Manufacturing Engineers did not voice any problems with the storage in the H-A. It is possible that this group simply had less need for storage. Alternatively, because this group also reported the highest levels of environmental control, it may be that the greater degree of control helped to reduce the need for storage.

10. Conclusion

***Control.*** One of the primary objectives of the H-A design was to provide users with environmental control over the workstation through adjustability of interior workstation features and ability to move the workstation itself. When compared to users of standard cubicles, H-A users reported significantly higher availability of control. That is, H-A users reported a higher proportion of adjustable workstation features than workers in standard cubicles. In addition, H-A users were found to have greater knowledge of control of their workstation features, meaning that, compared to users of standard cubicles, H-A users knew how to adjust more of their adjustable workstation features.

***Individual Performance and Group Effectiveness.*** A survey was used to assess individual performance and group effectiveness both before the installation of the new workstations (H-A or systems furniture cubicles) and three months after installation. The increased control that resulted from the design of the H-A workstation was found to have a positive influence on both individual performance and group effectiveness.

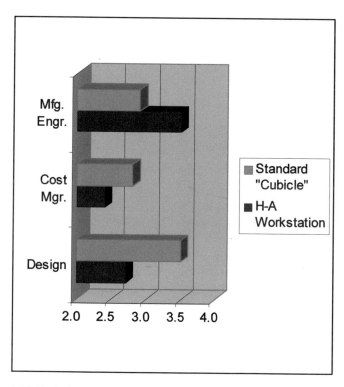

**Figure 6.13** Satisfaction with Storage (Author)

***Group Collaboration.*** Group collaboration was defined as the extent to which one's work group was supportive, participated evenly, and shared responsibility. When availability of control was greater, as was seen with users of the H-A, workers reported improved group collaboration.

***Stress and Physical Health Problems.*** Stress was measured according to how bothered, worried, frustrated, and tense one reported feeling at work. Findings from the present study indicated that, as knowledge of control increased, reports of stress decreased. Physical health problems were measured as reports of musculoskeletal discomfort. There were no significant findings that involved the reports of musculoskeletal discomfort.

## C. Summary

The hypothesis that the experimental H-A workstation would increase user control was supported. Both availability of control and knowledge of control were shown to increase significantly with the implementation of the H-A workstations. In addition, the hypothesis that there would be a positive relationship between the indices of control and individual performance, group effectiveness, and group collaboration was supported. Finally, our hypothesis that there would be a negative relationship between the indices of aspects of control and stress and musculoskeletal discomfort was partially supported.

Our overall model was supported and control had a favorable influence in every variable except for musculoskeletal discomfort. As such, it can be concluded that the design of the H-A workstation enables users a degree of physical control beyond that of standard workstations, thereby increasing performance, collaboration, and psychological health.

## V. CASE STUDY: SALES PROJECT TEAM WORK SPACES

### A. Purpose of Project

The purpose of this project was to test new workplace design concepts for sales team workers at the field sales office of a manufacturing and service company. In this study, we created open-office team spaces, a departure from that organization's practice of housing all employees in field sales facilities within private "drywall" offices. The success of our design concepts were tested at three levels: facility performance, work group performance, and individual performance. Facility performance was measured as a function of overhead costs and effective use of space (i.e., whether more employees could be comfortably accommodated in less space without negative reactions). Success of design concepts for team and individual workers were tested against the criteria of team participation and effectiveness, individual performance, psychological stress, and satisfaction with ergonomic aspects of the work space.

### B. Introduction

The present study analyzed worker perceptions before and after a move from traditional private offices to team work spaces specifically designed to enhance group effectiveness and individual performance. This study also assessed the impact of the test environment on worker reactions, including: control over the job, satisfaction with the environment, quality of lighting, privacy, control over the adjustability of the work space, and musculoskeletal problems.

To develop designs for the new space, we collected information by interviewing each employee. We then proposed two different design schemes to the employees. They unanimously selected an open-plan approach that used systems furniture to provide team spaces. Approximately four months after the move to the new space, a sound-masking system was installed to assess any changes to performance and satisfaction to the workers.

Data were collected from both the experimental and control groups at three points over the course of the study: prior to any changes, after the move to the redesigned facility, and after the installation of the sound-masking system.

### C. Problem

The analysis of the field sales office revealed problems with the work space at the facilities level, problems with support of work groups, and design and ergonomic problems at the level of the individual worker.

## 1. Facilities

The field sales offices require a substantial amount of square footage to house a given number of people. The layout of each office may be less efficient than possible and, because of the private office layout, may be a less than optimal solution to the company's need for flexibility in supporting changing staffing levels as internal needs change.

*Design Intervention.* After interviews with all employees, we developed an open-plan systems furniture-based design solution that substantially reduced the amount of space required to house all employees, and also provided the flexibility to quickly add staff without reducing the quality of the space.

## 2. Work Group

The sales people work in stable Project Team structures (see our discussion of team types earlier in this Chapter) as they strive to identify potential clients, develop sales proposals, and support existing clients. The sales process is complex, and a typical sales team may make from one to three sales per year. Members of sales teams require almost continuous face-to-face contact over the course of a typical working day. Prior to this study, all sales staff were housed in private offices arranged in a linear sequence that did not facilitate such interaction.

Technical support workers make up Work Teams (see our discussion of team types earlier in this chapter) and their group has overall responsibility for effectively supporting ongoing client technical needs, closely supporting the client's Systems Administration team. While their responsibilities are individual, the Technical Support group does have a unified mission and strong group identity. Currently Technical Support people do not have assigned office space because so much of their work is out of the office on client locations. This causes a problem in terms of lack of environmental support for their ongoing training, administrative tasks, group meetings and information sharing, and group morale.

*Design Intervention.* We designed team spaces to house the members of the sales/marketing team. These spaces provide individual workstations clustered around a common core area for group meetings. The team spaces are constructed from panels and use systems furnishings and a freestanding table for group meetings. The spaces were intended to provide flexibility for changes in organizational structure over time and to enhance individual interaction while maintaining privacy.

We designed a team space for the Technical Support group that provides technology support, space for computer based training, informal meeting space, and an identifiable group area with unassigned individual spaces.

## 3. Individual Workspace Problem

Our analyses of the existing environment revealed that many aspects of the work spaces, furnishings, lighting, and equipment layout did not meet ANSI/HFS 100 VDT Workstation ergonomic guidelines, increasing the possibility of work-related injuries due to the design of the work spaces and potentially reducing individual performance.

***Design Intervention.*** In the new test environment we designed work spaces to be ergonomically correct for the type of work performed within.

## D. Design Hypotheses

Three design hypotheses were developed from the interviews with employees and the subsequent problem identification described in the previous section.

## 1. Hypothesis 1: Facilities Performance

We hypothesized that the design intervention will reduce the square footage required to support the activities of the workers in the field sales office. This intervention will also maximize the flexibility of the facility by allowing additional staff to be added as the office grows without reducing existing work space size or causing perceptions of crowding.

## 2. Hypothesis 2: Work Team Support

We hypothesized that the experimental team space will significantly enhance group interaction and effectiveness.

## 3. Hypothesis 3: Individual Work

We hypothesized that the test space will significantly enhance the quality of work life for individual employees on a number of dimensions, including: individual performance, stress and health issues (such as musculoskeletal pain), and ergonomic issues such as adjustability of the work space, lighting, and visual and auditory privacy.

## E. Methods

### 1. Pre-design Programming and Design Development

Work space requirements for the sales team and clerical support workers were developed through a research-based process that involved individual interviews. Two design schemes for the new work spaces for the sales team were developed and presented to the entire office of participating employees. One design combined drywall offices for all employees with central team meeting spaces. This design was somewhat similar to the original office in which every employee was housed in a private office. The alternative design used an open-office concept with ergonomically designed team and temporal work spaces built using systems furniture. The employees unanimously voted to try the open-plan systems furniture approach.

### 2. Sites

This study examined three field sales offices: one was the location of the design intervention ("test" site), and two other field sales offices served as control groups against which measures of changes in worker reactions were assessed.

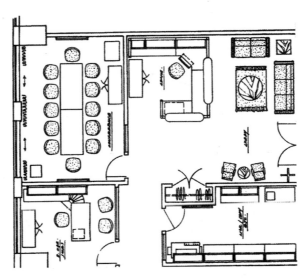

**Figure 6.14** Portion of floor plan of Field Sales office prior to redesign (Reprinted with permission from *A Test of Team and Temporal Work Spaces*. Copyright 1994 by Herman Miller, Inc. All rights reserved.)

*Interiors Prior to Intervention.* Figure 6.14 shows a schematic floor plan of the Field Sales Office prior to the experiment. Note that employees are housed in private offices. The floor plans for the control group sites are similar, using private offices to house employees.

Figure 6.15 shows the main entry area used by the Experimental Site prior to the redesign. The private offices are adjacent to it, and are used by members of the Sales Team. They were furnished with free-standing metal desks and individual storage and shelving (see Figure 6.16).

*Interiors after Design Change.* After the research process was complete, the consulting team then translated the data from the sales team members into a "needs set" that took into consideration workers' interaction needs, communication patterns, team structure, team goals, and business unit mission. We developed design criteria from this needs set that were then translated into design concepts. We generated two alternative design concepts for the office: one in which team spaces were defined through full-height drywall construction, and another in which an open, collaborative work environment with individual and group spaces was created using office systems furniture. The team chose to implement the latter of the two design alternatives.

Figure 6.17 shows the implementation of the mix of small private offices on the building perimeter and the adjacent individual and team meeting spaces for the Sales professionals. This option, the one in which team spaces were created through the use of office systems furniture, represented a great leap away from the norm for this organization.

## 3. Research Design

This research uses an "untreated control group design with pretest and posttest," which is a frequently used design in field research (Cook and Campbell, 1979). Figure 6.18 graphically depicts the design of this study ("O" = observations and "X" = the experimental interventions). The top half of the box shows the experimental group, and the bottom half shows the control groups.

The first observations were made on both groups simultaneously before any changes to the work environment were introduced (see Figure 6.18, "$O_1$"). Next, the new team work spaces were installed for the experimental group (see Figure 6.18, "$X_1$"). Approximately 1 month after the test group had been moved to the new team spaces, both groups were again observed (see Figure 6.18, "$O_2$"). After this, a sound-masking system was installed at the test site (see Figure 6.18, "$X_2$"). Approximately 1 month after the installation of the sound masking, final observations were made (see Figure 6.18, "$O_3$"). This research design makes it possible to ascertain whether any changes to the test group are due to the design interventions, and helps to rule out many other possible causes.

**Figure 6.15** Main entry area prior to redesign(Author)

**Figure 6.16** Typical office interior for professional worker (Author))

**Figure 6.17** Private office and group space concept (Author)

A total of 30 people participated in the study. Because we collected data from this group at three points over the course of the study, we obtained a total of 67 usable questionnaires. Data were collected through intensive one-on-one interviews and confidential, individual questionnaires.

$$O_1 \quad X_1 \quad O_2 \quad X_2 \quad O_3$$

$$\rule{5cm}{1pt}$$

$$O_1 \qquad\qquad O_2 \qquad\qquad O_3$$

$$\rule{5cm}{1pt}$$

$$O_1 \qquad\qquad O_2 \qquad\qquad O_3$$

**Figure 6.18** Design of field study (Author)

4. Measures

A questionnaire examined perceptions of: individual performance, group effectiveness, group participation, stress, adjustability of work space, lighting, privacy, work demands (pressure), job satisfaction, control over job, and musculoskeletal problems. Some of these questions have been developed and used in previous research (Brill et al., 1984; Caplan et al., 1975; Nadler, 1977; O'Neill, 1992).

Multiple-item measures of these constructs were created. The reliability of these measures was tested using Chronbach's alpha ($\alpha$), a statistic that allows researchers to ask a number of questions relating to the same concept (for example, privacy) and then statistically collapse the responses to these questions into a single index representing that construct. Chronbach's alpha can range from 0 (representing a completely unreliable index) to 1 (meaning a highly reliable and useful index). The greater the alpha value, the more reliable the index.

5. Performance Measures

***Group Effectiveness.*** This index is composed of four questionnaire items measuring the effectiveness of the group or team, including: problem solving, making decisions, getting the work done, and overall effectiveness ($\alpha = .93$). This scale uses a seven-point Likert-type scale where 1 = ineffective and 7 = effective. Employees not specifically members of a recognized team were asked to complete these questions with regard to the effectiveness of all sales office employees as a team or group.

***Group Participation.*** This index is composed of four questionnaire items measuring the participation of individuals within the life of the team, including: feeling like a part of the group, involved in activities, even participation by members, and supportiveness of team members ($\alpha = .94$). This scale uses a seven-point Likert-type scale where the value labels vary by question. In general 1 = a lower amount of the attribute and 7 = a higher amount of the attribute.

***Individual Performance.*** This index is composed of nine questionnaire items, including measures of: amount and quality of work accomplished, meeting deadlines, interpersonal relationships, sense of responsibility, and creativity ($\alpha = .94$). This scale uses a nine-point Likert-type scale where 1 = unacceptable performance and 9 = perfect/ideal performance.

6. Health Issues

*Stress.* This eight-item index assesses short-term stress reactions in workers during the past month, including: heart rate, breathing, appetite, dizziness, stomach problems, and trembling hands ($\alpha = .74$). This scale uses a 3-point Likert-type scale where 1 = never and 3 = three or more times (Caplan et al., 1975).

*Muscle Pain.* This five-item index assess the amount of upper body pain experienced by the respondent, including measures of: neck pain, headaches, and shoulder and arm pain ($\alpha = .88$). This scale uses a 5-point Likert-type scale where 1 = never and 5 = always.

7. Work Space Ergonomics Issues

*Flexibility.* This six-item index assesses the adjustability and flexibility of chair, furnishings, lighting, and work materials ($\alpha = .76$). This index provides information about the amount of control employees have over their immediate work environment. This scale uses a 5-point Likert-type scale where 1 = very dissatisfied and 5 = very satisfied.

*Visual Privacy.* This index is made up of three variables related to visual privacy, including: distractions resulting from seeing others, interruptions, and being exposed to the view of others. This scale uses a 5-point Likert-type scale where 1 = strongly disagree and 5 = strongly agree. The items were reversed for purposes of statistical analysis; like the other scales, a higher value indicates more of the condition (more privacy).

*Auditory Privacy.* This index is made up of three items relating to the ability to have confidential conversations, overhearing others, and being overheard. This scale uses a 5-point Likert-type scale where 1 = strongly disagree and 5 = strongly agree. These items were also reversed for purposes of statistical analysis; like the other scales, a higher value indicates more of the condition (more privacy).

*Glare Problems.* This is a single questionnaire item that assesses the amount of glare caused by the lighting at the workstation. This question uses a 3-point Likert-type scale where 1 = great deal of glare and 3 = doesn't cause glare.

## F. Results

This section describes in detail the findings related to each of three research hypotheses predicting that redesign of the space to support teamwork using an open-plan systems furniture approach would significantly enhance facility performance, group effectiveness, individual performance, and health. To simplify this Case Study, we combined the responses of both Sales and Technical Support Team members in our analysis of responses.

## 1. Facilities Performance

***Real Estate Costs.*** The test design significantly reduced square footage requirements for a typical field sales office while increasing the number of employees that can be accommodated. In this test, the per-person square footage requirements were reduced from 261 in the old space to 169 square feet in the new (a 35 percent reduction) even though the amount of support space increased and the number of employees accommodated increased from 14 to 23. This flexibility in accommodating growth (or reduction) in staff levels also enhances corporate control over facilities.

## 2. Work Group Issues

***Group Collaboration.*** Figure 6.19 shows comparisons on self-rated group participation for workers who received the redesigned work environment (Experimental group) and the Control group, who remained in private offices. Group collaboration is a construct that measures the level of individual participation within the life of the team or group, including: involvement in activities, supportiveness of team members, and other aspects of individual involvement in team life.

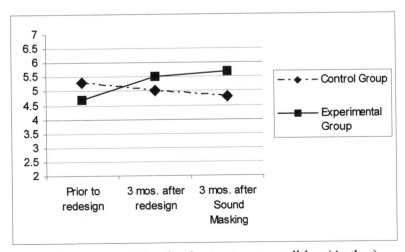

**Figure 6.19** Group participation by treatment condition (Author)

At the first point of measurement, "prior to redesign" workers in the Experimental group, whose spaces had not yet been redesigned, experienced significantly lower levels of group participation than did workers in the Control group at the other two locations. After this initial measurement was taken, the Experimental group was relocated to their new office environment, which had been redesigned specifically to support group and individual collaboration. To avoid unintended effects on the outcome measures (the "Hawthorne effect") due to workers' feelings about the move or participation in the study, we waited 3 months after the move was complete before measuring the Experimental and Control groups again. We felt that this was sufficient time for workers to become acclimated to their new location and re-establish work routines.

At this second point of measurement, we found that group participation was approximately 4.5 percent higher for workers in the Experimental work environment when compared to the workers in the Control groups at the other two locations. Another way of examining changes in group participation is to track changes in the Experimental group before and after the move. In this case, the Experimental group acts as its own Control group, since we measured group participation in this group before and then after anything was done to their work environment. When we examined the difference in group participation before and after the move for the Experimental group, the change in group participation was even more pronounced, increasing 15.7 percent after the move to the redesigned environment.

Shortly after the second measurement point, a sound-masking system was installed in the Experimental work environment. Three months after the installation of that system group participation was measured a third time. At this point, group participation levels for the Experimental group were significantly greater than for the workers in the control groups. Thus, the increase in group participation in the Experimental group from its levels before the redesign were largely sustained. These findings suggest that it is possible to create work environments that are not totally reliant on traditional tactics of closed offices and conference rooms, and still improve group cohesiveness.

*Group Effectiveness.* In this study, we measured group effectiveness as a function of problem solving, getting work done, and appraisal of group overall effectiveness. Figure 6.20 shows that between the first two measurement points, both the Control and Experimental groups had similar, slightly downward trending, scores. However, at the final measurement point, the Experimental group showed a significant increase in self-assessed group effectiveness, when compared to workers in the Control group. Evaluations of group effectiveness tend to remain stable over time. We believe that this type of finding reflects the systemic nature of the work environment, in that there are rarely direct, cause-and-effect relationships between the environment and outcomes such as performance. Rather, performance effects are moderated by a number of variables (technology, group design, job design, management styles, etc.) of which the physical environment is one variable.

***Individual Performance.*** Individual performance was measured as a function of quantity and timeliness of work accomplished, and other measures of personal responsibility for work and internal relationships. The results for our tracking of individual performance are shown in Figure 6.21. Figure 6.21 shows comparisons on self-rated individual performance (rated on a 9 point scale) for workers who received the redesigned work environment (Experimental group) and the Control group, who remained in private offices. There was no statistically significant difference in individual performance between workers in the Control group and Experimental group before the move to the redesigned facility (Mean of Control Group = 6.14, Mean of Experimental Group = 6.08).

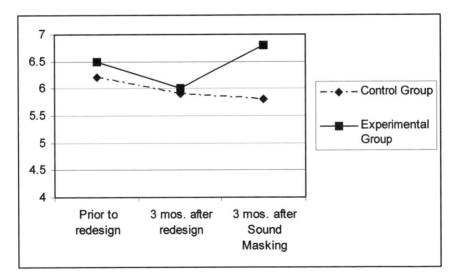

**Figure 6.20** Group effectiveness by treatment condition (Author)

This shows that all workers participating in the study began with roughly the same level of individual performance. After this initial measurement was taken, the Experimental group was relocated to their new office environment, which had been re-designed specifically to support group member collaboration. To avoid unintended effects on the outcome measures (the "Hawthorne effect") due to workers' feelings about the move or participation in the study, we waited three months after the move was complete before measuring the Experimental and Control groups again. We felt that this was sufficient time for workers to become acclimated to their new location and re-establish work routines.

At this second point of measurement, we found that individual performance was significantly greater for workers in the Experimental work environment when compared to the workers in the Control groups at the other two locations. Thus, although the individual performance scores in both the Control and Experimental groups rose in tandem over time, the workers in the Experimental group reported significantly greater levels of individual performance.

Shortly after the second measurement point, a sound-masking system was installed in the Experimental work environment. Three months after the installation of that system, group participation was measured a third time. At that point individual performance levels between the Experimental group and Control groups diverged, with workers in the Experimental work environment reporting significantly greater individual performance than workers in the Control groups.

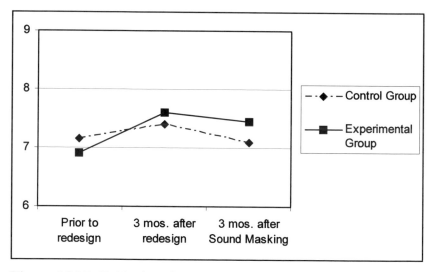

**Figure 6.21** Individual performance by treatment condition (Author)

In support of our design hypotheses, the final performance measurement revealed that the move to the redesigned test site resulted in a significant, sustained increase in the individual performance of workers within the experimental offices when compared to workers at the two control sites. In addition, individual performance of workers within the experimental work environment increased by approximately 5 percent when compared to their performance within their original private offices before the intervention began. We believe that the increase in individual performance at the final time measurement is due in part to the quality of the redesigned workstations (which the results in subsequent sections of this report show had superior design and ergonomic qualities) and the implementation of the sound-masking system, which may serve to more directly support the needs of individual workers engaged in solo tasks that require concentration, especially important to workers in open office, systems furniture based environments.

## 3. Work Space Design Issues

*Adjustability.* Adjustability of the work environment (ability to rearrange furnishings, equipment, lighting, and work materials) is related to the amount of control an employee has over the immediate work space. Environmental control is related to performance and satisfaction (O'Neill, 1992; 1993), and is a key premise underlying the discussion of effective work in this volume.

Baseline measures of availability of control (through adjustability) show that workers in the Control groups reported that their work environments gave them slightly more control than workers in the Experimental group before the redesign (see Figure 6.22).

Subsequent to the move to the new location, workers in the Experimental site reported a statistically significant, 30 percent increase in control within their work spaces as compared to workers in the Control groups at the same time period.

This improvement is probably due to the ergonomically enhanced furniture that replaced the metal desks used by the test participants in their original work environment. At final measurement time, perceptions of control had increased for both Experimental and Control groups, probably in part due to heightened awareness of this issue from answering the questionnaire, and learning about features of the work environment from interacting with it. We do not believe that the addition of sound masking in the experimental environment was related to the increase of perceived control in the experimental group. Workers in the Experimental group maintained a significantly greater sense of control over the workstation than did workers in the Control groups at this final measurement point.

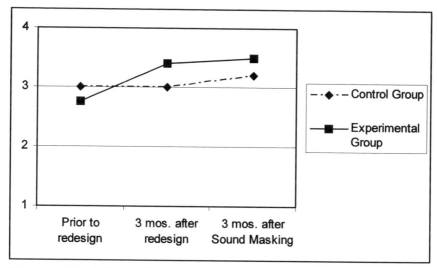

**Figure 6.22** Adjustability of workstation features by treatment groups (Author)

***Auditory Privacy.*** Auditory privacy is made up of three items: ability to have confidential conversations, overhearing others, and being overheard. Like the other scales, a higher score indicates greater privacy (see Figure 6.23). This Figure shows that the Control groups enjoyed satisfactory levels of auditory privacy throughout the course of the study. Figure 6.23 shows that at the first measurement point in the study ("Prior to Redesign"), the Experimental group had somewhat lower levels of auditory privacy. After the move to the new, open-plan space ("3 Mos. after Redesign"), sound privacy for workers in the experimental location dropped significantly below levels enjoyed by their counterparts in the control groups, which were traditional private offices.

With the addition of sound masking later in the study, (see Figure 6.23, "3 mos. after Sound Masking") auditory privacy levels in the Experimental site rose to levels reported by workers in the Control sites in the first two data collection times in the study, but continued to lag behind sound privacy levels reported at the third data collection point. This result points out the importance of providing sound masking in conjunction with a more visually open environment to preserve perceptions of privacy, especially when employees have previously been accustomed to private offices. Sound masking can be economical, especially if installed during the initial "build-out" of the space.

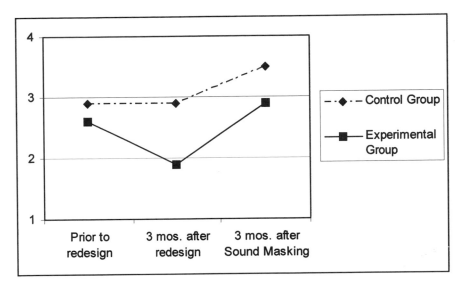

**Figure 6.23** Auditory privacy by treatment groups (Author)

*Visual Privacy.* Visual privacy was measured through three questionnaire items: distractions from seeing others, interruptions, and being exposed to the view of others. Visual privacy was reported on a 1 to 5 scale (5 being the greatest amount of privacy). Figure 6.24 shows that the Control groups reported stable, satisfactory levels of visual privacy throughout the course of the study. Figure 6.24 also shows that at the first measurement point in the study ("Prior to Redesign") the Experimental group had somewhat lower levels of visual privacy than the Control groups, but still were at a satisfactory level (Mean score = 3.0). After the move to the new, open-plan space ("3 mos. after Redesign"), visual privacy for workers in the Experimental location dropped significantly below levels enjoyed by their counterparts in the Control groups, who were housed in traditional private offices.

With the addition of sound masking later in the study, visual privacy levels in the Experimental site rose to levels reported by workers in the Experimental site prior to the redesign (see Figure 6.24) and almost to levels reported by the Control group at the final data collection point. We believe that auditory and visual privacy are closely linked constructs, and, in fact, that the addition of sound masking did affect perceptions of visual privacy. It is also possible that workers in the Experimental group learned techniques for preserving privacy over time that helped to compensate for the relatively more open spaces they found themselves working within, which explains in part the rise in perceptions of visual privacy over time.

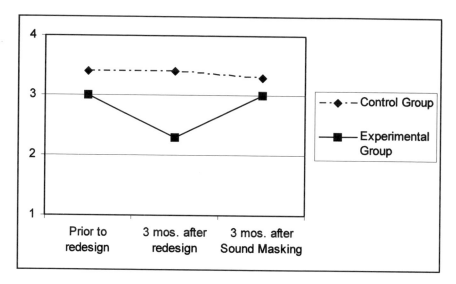

**Figure 6.24** Visual privacy by treatment groups (Author)

*Glare.* Quality of lighting in the work space has been linked to performance and health in offices. Glare on the computer screen is a common problem in offices. We assessed the level of existing glare by a single question, using a three-point scale in which 1 = no glare present on CRT screen, and 3 = great deal of glare. Figure 6.25 shows that glare levels in the control groups remained constant over the course of the study.

This is logical since no changes in the design of the physical design of offices were made to those groups during the study time period. The results show that prior to the redesign, workers in the Experimental group experienced significantly greater problems with glare than did workers in the Control groups (see Figure 6.25). After the redesign of the work environment, workers in the test site experienced a significant, 31 percent decrease in reported glare problems caused by lighting in their work spaces. This is especially significant because the test site had floor-to-ceiling glass curtain walls, as compared to their original private offices which had only standard size windows. Thus the potential for glare problems was greater due to the building design of the test site. Because of this factor, we specifically designed the work space layout in a manner that we hoped would limit glare problems. This was possible because of the greater flexibility afforded in layout planning through the use of systems furnishings.

By the third data collection point, reported glare problems rose significantly above what had been reported at the second data collection point, but were still within the acceptable levels initially reported by the control groups (see Figure 6.25). We believe this rise in reported glare problems was due to seasonal changes in length of day and position of the sun during working hours. This finding underscores the need to collect information about work environments longitudinally so that temporal effects can be considered. Thus, by the third data collection point, glare problems within the Experimental, open-plan site were essentially the same as those in the private office Control groups.

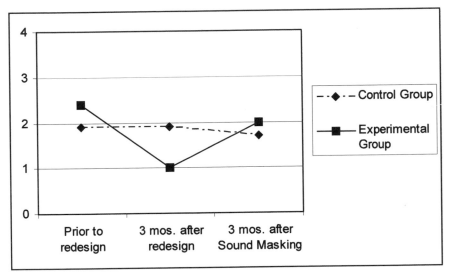

**Figure 6.25** Glare problems by treatment group (Author)

## 4. Health Issues

***Psychological Stress.*** We measured stress through an 8-item index that assessed short term, psychosomatic stress reactions. Using a 3-point scale, we asked workers how often they had experienced stress symptoms in the last month. On this scale 1 = never experience the stress symptom, and 3 = three or more times in the last month. Figure 6.26 shows that at the start of this study, prior to any changes in the work environment, workers in the Experimental group reported significantly higher stress levels than did workers in the Control groups.

This study took place during a time when the participating company reduced its work force by 9 percent as part of a major restructuring. Over the duration of this study, all employees in the Experimental group reported feeling significantly increased work pressure (feeling overworked, pushed by deadlines).

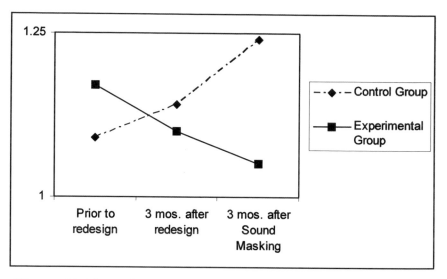

**Figure 6.26** Stress levels by treatment groups (Author)

By the time of the second data collection, three months after the redesign of the office space, reported stress of workers in the Experimental group had dropped to about the level reported by workers in the Control groups. Note that the reported stress level of workers in the control group is on a continuing rising trend throughout the duration of the study. By the third data collection point, three months after the installation of the sound masking, workers in the Control groups reported the highest levels of stress recorded in the study, while workers in the Experimental offices reported the lowest stress levels at any time (see Figure 6.26). This is an important finding, since stress reactions have been linked to employee burnout and serious health problems.

While we assume that all employees in the company were experiencing somewhat equivalent stress levels due to the restructuring occurring within the organization, we believe that, in some part, the enhanced work environment, including the sound masking, may be a causal factor in the reduced stress levels reported here.

***Muscle Pain.*** We measured upper body pain through a 5-question index that included assessments of neck pain, headaches, and shoulder and arm pain. This index used a scale in which 1 = never experience the pain and 5 = always experience the pain. Figure 6.27 shows that upper body pain remained relatively constant for workers in the Control groups over the course of the study.

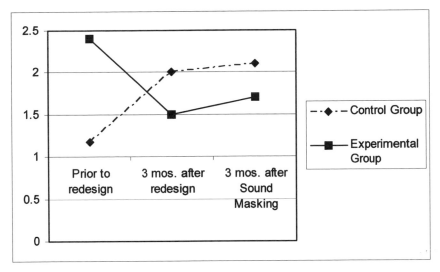

**Figure 6.27** Upper body pain by treatment group (Author)

Figure 6.27 shows that the Experimental group reported significantly greater levels of upper body pain than the Control group prior to the redesign of their work environment. By the second data collection point, 3 months after the redesign of their work environment, workers in the Experimental work environment experienced a significant reduction in reported upper body pain as compared to the Control groups. We believe that much of this improvement is due to the use of ergonomically appropriate furnishings in the test space. Note that even after the final data collection point, these substantial reductions in upper body pain were sustained, along with the concurrent reductions in psychological stress. Upper body pain is, in part, a long-term health reaction to poor ergonomic working conditions; along with untreated psychological stress, it can reduce the effectiveness of the organization through loss of work and health insurance claims.

## VI. CASE STUDY: WORK TEAM SPACES FOR
## INSURANCE CLAIMS PROCESSORS

### A. THE PROJECT

Many organizations are seeking ways to use their human capital more effectively. Even work that had been almost the equivalent of rote assembly labor (such as that done by telecommunications operators or data entry workers) is being redesigned, giving workers a higher degree of autonomy and responsibility for their work. These types of jobs, which were once designed around the individual, are now being designed around teams. Work teams are being given levels of responsibility and authority to control the flow of their work and delegation of tasks to team members that was formerly the province of managers.

In addition, teamwork and group performance have become important issues for businesses that rely on collaborative efforts to meet organizational goals. These trends have put pressure on organizations to provide supportive work environments to meet the needs of work teams to leverage their efforts to the greatest extent possible.

In the present Case Study, we analyzed the newly created Work Team job design of an insurance claims processing department. Prior to our intervention, these workers were performing their jobs within workstations that had been designed for individual jobs. We were asked to create a project in which we would realign the work environment with the new unit of work, which was the Work Team. In this study, we selected one self-directed Work Team to use as a pilot project. Our goal was to redesign the physical space to support this new way of working.

### 1. Description of Existing Work Environment

The facility is located in the Midwest of the U.S. The design intervention was performed within the claims processing group. The existing workstations were constructed of standard systems furniture with panel-hung work surfaces and some individual and group storage. Each workstation contained two computers, including one networked PC that was used to support the main task, data entry, and a second "thin client" terminal that was used to access mainframe applications and data (see Figure 6.28). The seating used by the operators was height adjustable (by turning the seat pan in one direction or another). The seating was over 10 years old and worn out.

**Figure 6.28** Typical workstation prior to redesign (Author)

## 2. Structure of the Work Team

The participants in this case study were a team of four female data entry clerks who process insurance claims. They also update information on existing policies. They constitute a Work Team in the sense that the overall workload is evenly shared. This team performs regular, ongoing work. The members of this Work Team have shared performance goals. While not formally a "self-managed" team, this team made many of the decisions that would have been the responsibility of a manager in terms of allocating their internal resources and managing workload within the group.

## 3. Work Process

The tasks performed by the team are highly repetitive and structured. As new claims files enter the queue for processing, they are placed on shelves within the operators' workstations, or on shelving units located adjacent to the cluster of four workstations housing the team. When an operator is ready to process more claims, the operator walks to the central shelf or to another team member's workstation and retrieves more forms.

## 4. Analysis

There are several ergonomic design problems related to individual and team process that we identified in our analysis of the claim form processor's workstations. The work surfaces were not height adjustable and were at an incorrect height for the work being performed. Work materials that are in process are placed in front of the dumb terminal, which effectively blocks the use of that terminal, and occasionally fall on the floor. The operator must twist to access the forms, and there is no work surface or tool to support the display of the paper form while the operator is processing it. When retrieving more forms to process, the operator has to hunt around co-workers' workstations or the central storage area to find forms. This activity wasted time and created extra bending and potential for awkward postures.

## B. Design Intervention

Our design intervention consisted of a redesign of the four workstations used by this claims processing Work Team to better support individual and group work process. A team space made up of four workstations was designed using an integrated, vertical storage and display unit that served as one side of each workstation, and furniture systems products (see Figure 6.29). In addition to this furniture, these participants also received low-back task chairs, because the chairs they were using prior to the test were broken and worn out.

Figure 6.30 show additional details of the interior spaces of the individual workstations. Figure 6.30 shows the vertical, integrated storage and display unit that served as one side of each workstation. This storage unit was where claims files were stored while awaiting processing. This display serves as a visual indicator of the pending workload for each team member, and permits other team members to assess the overall work flow within the team and reallocate workload or tasks as needed to meet the group performance goals.

Note the hanging work tools that serve to hold claims files that are in various stages of processing, and to keep the work surface clear for ongoing work. In addition, the work space was designed to support the use of two VDT terminals in conjunction with the paper-based tasks.

## C. Methods

### 1. Research Design

This is a longitudinal study in which user perceptions were measured at three points in time over the duration of the project. At the site, the perceptions of a Control group of employees in the same company, performing identical work in individual workstation "cubicles" were also assessed.

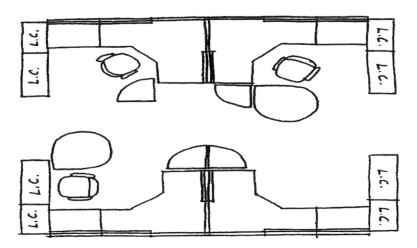

**Figure 6.29** Layout of Redesigned Group Work Environment (Reprinted with permission from Research report: LiaisonTM User Test. Copyright 1993 by Herman Miller, Inc., Product Research Group. All rights reserved.)

**Figure 6.30** Workstation with Vertical Integrated Storage/Display Unit (Author)

The first data collection occurred 2 months before the product was installed. The second data collection occurred about 1 month after installation of the test product. The final data collection point was 4 months after the test product was installed. The research used an "untreated control group design with pretest and posttest" which is used frequently in field research (Campbell and Stanley, 1966). See Figure 6.31 for a graphic depiction of the study design, (O = observations and X = introduction of the Experimental team). The first observations were done on both groups simultaneously before any changes to the work environment were introduced (O1).

| Experimental Group | $O_1$ | $X_1$ | $O_2$ |
|---|---|---|---|
| Control Group | $O_1$ | | $O_2$ |

**Figure 6.31** Untreated control group design with pretest and posttest (Author)

Use of control group data helps reduce the threats to the internal validity of the study, or the effects of outside influences on test results (such as organizational changes in the company or the national economy).

## 2. Participants

The study participants consisted of 15 claims processing employees at the insurance company. Four of the employees were in the Experimental group, the balance were in the Control group.

## 3. Measures

A 20-minute, self-administered questionnaire was distributed through a contact person at the site and was completed by employees during regular working hours while seated in their workspaces.

The questionnaire examined perceptions of: control over pace of work, stress, musculoskeletal complaints, individual performance, group effectiveness, and privacy. Evaluations of the workspace were also recorded, including: overall satisfaction, quality of storage, degree to which workspace aids efficiency and organization of materials, and flexibility/control over workspace. Some of these questions have been developed and used in previous research (Brill, 1984; Caplan et al., 1975; Nadler, 1977; O'Neill, 1992).

Multiple-item measures of these constructs were created and the reliability of these measures tested using Chronbach's alpha ($\alpha$). This statistic allows researchers to ask a number of questions relating to the same concept (for example, privacy) and then statistically collapse the responses to these questions into a single index representing that construct.

Chronbach's alpha ($\alpha$) can range from 0 (representing a completely unreliable index) to 1 (meaning a highly reliable and useful index). Thus, the greater the alpha, the more reliable the scale.

*Individual Performance.* Individual performance was measured through eight items (each using a 9-item Likert-type scale). The scale asked people to evaluate how well they do their job within a range from 1 = unacceptable, 5 = reasonably well, to 9 = perfect/ideal. The individual ratings included: amount of work accomplished, quality of work, error rate, taking responsibility, creativity, getting along with others, dependability, and overall performance (Brill, 1984). These items were averaged to form an index ($\alpha$ = .86).

*Group Effectiveness.* Each participant was asked to rate the effectiveness of their work group on seven dimensions (each using a 7-item Likert-type scale). On this rating scale, 1 = ineffective and 7 = effective. The following items were averaged to form an index ($\alpha$ = .91): problem solving; making decisions; getting work done; using member skills, abilities, and resources; meeting individual needs; satisfaction with group membership (the scale anchors for this item are 1 = very dissatisfied to 7 = very satisfied); and overall effectiveness of group (Nadler, 1977).

*Stress.* To measure psychosomatic symptoms of stress, eight items measuring this construct (on a 3-point Likert scale) were averaged to form an index ($\alpha$ = .69): heart beating hard, shortness of breath, dizziness, hands trembling, upset stomach, heart beating fast, ill health, and loss of appetite (Caplan et al., 1975).

*Workspace Aids Efficiency.* Each participant was asked to rate the degree to which their workspace enhances their efficiency. To measure workspace efficiency, 4 items measuring this construct were averaged to form an index ($\alpha$ = .61): workspace helps me get my job done efficiently; finding work materials quickly; repeating tasks unnecessarily (reverse scored); and losing track of where I am in a task (reverse scored).

*Quality of Storage.* To measure storage quality, three items measuring this construct (on a 5-point Likert scale) were averaged to form an index (Pearson's r = .57 to .74 between variables): satisfaction with amount of storage, number of file drawers, and number of shelves.

*Stretching and Twisting.* To measure problems with users having to stretch and twist while reaching for work materials within the workspace, six items measuring this construct (on a 5-point Likert scale) were averaged to form an index ($\alpha$ = .87): bothered by having to stretch or twist, working in uncomfortable positions, awkward work motions, stretching or twisting to reach work materials, having to reach too far while seated, and having to twist too far while seated.

*Privacy.* Separate scales were developed for visual and auditory privacy. Three items measuring auditory privacy were developed (using a 5-point

Likert-type scale) and averaged to form an index (Pearson's r = .48 to .86 between variables): ability to have confidential conversations in workspace (reversed), overhearing immediate neighbors, and being overheard from the workspace. Four items measuring visual privacy (on a 5-point Likert-type scale) were averaged to form an index ($\alpha$ = .80): distracted by seeing others, interrupted while working, too exposed to view of others, and satisfaction with privacy (reversed).

*Control Over Organization of Workspace (Environmental Control).* Each participant was asked to rate the degree to which the workspace supports the rearranging and organization of work. Six items measuring this construct (on a 5-point Likert scale) were averaged to form an index ($\alpha$ = .92): ability to arrange work materials to match work style, satisfaction with arrangement of equipment, ease of organizing work materials, ability to rearrange work materials to suit the task at hand, and amount of flexibility/adjustability in workspace.

*Control Over Pace of Work.* Control over pace of work was measured by a single item (measured on a 5-point Likert-type scale): "feeling in control of the pace of my work."

*Satisfaction with Workspace.* This variable was measured through a single item (measured on a 5-point Likert-type scale): "In general, I am satisfied with my workspace."

*Musculoskeletal Pain.* Five aspects of musculoskeletal pain were measured on a 3-point Likert-type scale (1 = never, 3 = often), including: back pain, pain in arms and hands and fingers, pain in neck and shoulders, pain in legs, and general muscle pain. These indexes were created from questionnaire items developed by Sainfort (1990).

*Visual Problems.* Visual problems were assessed through items measured on a 3-point Likert-type scale (1 = never, 3 = often). These items assessed frequency of headaches, burning eyes, tearing or itching eyes, and eye strain or sore eyes. These indexes were created from questionnaire items developed by Sainfort (1990).

## D. Analyses

### 1. T.-tests

The first set of analyses for each site consists of t-tests computed to discover differences between the reported means on the outcome measures (for example, on privacy and performance) between the Control and Experimental groups. This analysis determines statistically significant differences between the groups on the outcome measures but does not indicate the cause of the differences.

Participants using the Experimental workstations spent significantly more time in their workspaces than did the Control group (Control M = 79.7 percent of workday, Experimental M = 94.0 percent of workday, t = 3.30, df = 1, 38, p < .01) (see Table 6.03).

Participants in the Experimental workstations reported significantly higher ratings on workspace efficiency (workspace helps me get my job done efficiently; finding work materials quickly; not repeating tasks unnecessarily; not losing track of where I am in a task) (Control M = 3.87, Experimental M = 4.41, t = 2.75, df = 1, 39, p < .01). (see Table 6.03)

The Experimental group reported significantly fewer problems with auditory privacy (Control M = 4.16, Experimental M = 2.88, t = 3.54, df = 1, 40, p <.01); fewer problems with visual privacy (Control M = 2.9, Experimental M = 1.94, t = 2.9, df = 1, 40, p <.001); greater control over the pace of their work (Control M = 3.36, Experimental M = 4.33, t = 3.60, df = 1, 40, p <.05); and significantly greater control over rearrangement and organization of work materials, equipment, and furnishings than workers using standard furniture in the control group (Control M = 3.51, Experimental M = 4.12, t = 3.21, df = 1, 38, p <.01) (see Table 6.03).

## 2. Multiple Regressions

We followed the t-test analyses by multiple regressions, computed to determine the design features or other variables that may predict user responses. (For example, if we find significant differences in privacy between the two groups, what design features of the workspace are predictors of privacy? Through multiple regressions, we may be able to determine, for instance, that panel height is the greatest predictor of privacy. Thus, in this case, we would know that the difference in privacy between the two groups was related to panel height.)

The predictive effects of many of the study variables were examined. To examine the overall effects of the Experimental environment, employees from the control group were assigned a score of 0 and participants from the Experimental group were assigned a score of 1. This new variable was used as a "dummy variable" in the regression analyses.

Multiple regression information is presented in Table 6.04. This table uses several headings. The "R2" heading indicates the total amount of variance in the outcome measure predicted by all predictor variables.

The size of the numbers under the "Standardized Coefficient" heading indicates the relative size of the contribution of each predictor variable to the variance in the outcome measure.

| Table 6.03 | | | | |
|---|---|---|---|---|
| Significant Differences Between Control and Experimental Groups | | | | |
|  | **Control Group Mean** | **Experimental Group Mean** | **t value** | **Significance** |
| **% Time in Workspace** | 79.7 | 94.0 | 3.30 | p < .01 |
| **Efficiency of workspace** | 3.87 | 4.41 | 2.75 | p < .01 |
| **Auditory Privacy** | 4.16 | 2.88 | 3.54 | p < .01 |
| **Visual Privacy** | 2.90 | 1.94 | 2.90 | p < .001 |
| **Control Work Pace** | 3.36 | 4.33 | 3.60 | p <. 05 |
| **Control over Organization of workspace** | 3.51 | 4.12 | 3.21 | p < .01 |

The "T" heading is the t value for that variable. A higher t value indicates a greater level of statistical significance for the predictor variable. The "p" heading is the probability of error (chances in 100) in stating that the relationship between the two variables is statistically significant. Thus, if p =.02, there is a two percent chance that the relationship is not really statistically significant.

Hierarchical regression analyses were computed to determine the predictor variables for group effectiveness, individual performance, stress, visual privacy, and control over organization of work materials. These analyses show the relative contributions of each predictor variable on employee reactions. We found that assessed quality of storage was the most powerful predictor of group effectiveness, followed by use of the vertical storage element. Together these variables account for 21 percent of the variance in group effectiveness ratings. This finding suggests that appropriately designed storage and display space could enhance group effectiveness for clerical-type workers that work in teams.

The degree to which the workspace supports efficiency is the primary predictor of individual performance, followed by use of the vertical storage element. Together these variables account for 16 percent of the variance in self-evaluated individual performance. These findings suggest that a workspace that aids efficiency by helping workers keep track of work, coupled with use of a vertical storage and display element, could enhance individual performance for Work Teams performing this type of work.

We found that the extent to which the workspace supports efficiency, and quality of storage, about equally contribute to understanding 65 percent of the variance in reported control over the organization of work materials and furnishings in the workspace.

Visual privacy is the sole predictor of stress, accounting for 10 percent of the variance in stress scores.

The use of the vertical storage element predicts 13 percent of the variance in level of visual privacy experienced by these workers.

Because the vertical storage element predicts a significant amount of privacy, and privacy was significantly enhanced in the experimental group, the regression results suggest that use of the vertical storage element is related to a small but statistically significant reduction in reported stress levels of clerical workers in this Case Study. That relationship looks like the model depicted in Figure 6.32:

Experimental Workstation => Enhanced Privacy => Reduced Stress

**Figure 6.32** Model of Stress Predictors (Author)

We found that auditory privacy is predicted by three variables: visual privacy, quality of storage, and use of vertical storage elements. It is not surprising that visual privacy predicts auditory privacy, since they are both dependent on the amount of enclosure provided by the environment. Quality of storage may predict auditory privacy because high quality storage may permit employees to work for longer periods without leaving the workspace and thus lead to fewer overall distractions from noise in the workplace. The final predictor variable, use of vertical storage elements, may enhance auditory privacy because of the additional depth of the cabinets, which provide enclosure.

## E. Summary

At the beginning of this Case Study, four research hypotheses were stated predicting a relationship between the experimental work environment and improvements in individual productivity and group effectiveness, workspace satisfaction, privacy, organization of materials, sense of control over pace of work, ergonomics, and reduced stress.

Participants using the Experimental workstations spent significantly more time in their workspaces than did the Control group. Workers in the Experimental workstations reported significantly higher ratings on workspace efficiency (workspace helps me get my job done efficiently; finding work materials quickly; not repeating tasks unnecessarily; not losing track of where I am in a task). The Experimental group reported significantly fewer problems with auditory privacy; fewer problems with visual privacy; greater control over the pace of their work; and significantly greater control over rearrangement and organization of work materials, equipment, and furnishings than workers using standard furniture in the Control group.

## VII. CASE STUDY: WORK SPACES TO ENHANCE COLLABORATION AND SENSE OF COMMUNITY IN PROFESSIONAL JOB TYPES

### A. Study Background

A company in the consumer foods industry was experimenting with workspaces designed to enhance collaboration and communication between professional white collar employees. The company was attempting to improve collaboration and idea-sharing with these key employees, and also use the work environment to retain these key performers by enhancing sense of community and belonging. We conducted a Workplace Evaluation Study in order to gather information from their employees about their perceptions of this prototype work environment.

The intent of the study was to determine the effectiveness of this new space on enhancing collaboration, work group process, communication, and sense of community and belonging to the organization. We measured how employees felt about their workplace at two points in time; at their existing workspace before a move to a new, open work environment consisting of open systems furniture workstations (pre-move), and then also after the move to the new work environment (post-move).

### B. Workplace Issues Measured

To conduct the pre-move and post-move surveys, we used a Web-based survey system with 30 survey questions, to gather workplace data in five areas, including: Workplace (Attributes) Evaluation (lighting, storage, etc.), Collaboration, Work Group Issues, Communication, and the degree to which the workspace Communicates Community and Belonging. The survey also included "open-ended" comment fields within each of the five areas.

### C. Research Design

This was a field study using a pre-move and post-move measures design with a Control group. This is a commonly used research design that permits an assessment of the effects of workplace design on outcome measures, while controlling for extraneous effects (such as organizational or social changes, etc. that could affect participants' responses (Shadish, Cook and Campbell, 2001). This type of study design has been frequently used in a variety of office work research studies (O'Neill, 1998). Study participants were placed into two groups, which included a Control group, or employees that did not move to the new work environment, and an Experimental group, consisting of employees who did move to the new work environment. The team gathered data from both the Control and Experimental groups for the study using a workplace survey that was issued electronically to a total of 88 employees, consisting of 18 different work groups, from two company facilities. Figure 6.33 and the left photograph of Figure 6.34 show a typical configuration of the spaces used by employees before they moved to the new space, and of employees in the Control group. Figure 6.33 and the right photograph of Figures 6.35 and 6.37, show typical configurations for the new workspaces that employees within the Experimental group received.

**Figure 6.33** Typical Configuration of pre-move and Control group work spaces (Author)

## D. Results

1. Response Rate

The initial pre-move survey was distributed in Fall, and the post-move survey was distributed 4 months after move-in was complete. Overall, the pre-move survey response rate was 76 percent and 64 percent for the Control and Experimental groups, respectively. The post-move survey response rate was 81 percent and 75 percent for the Control and Experimental groups, respectively.

## 2. Workspace Evaluation

Overall, we found largely positive effects for the new work environment on employee perceptions, based on comparisons of the employees' assessment of the workplace before the move. In the Workspace Evaluation section of the survey, we asked employees to evaluate their work environment on nine workplace attributes. Employees were provided with 13 questions related to the following workplace attributes:

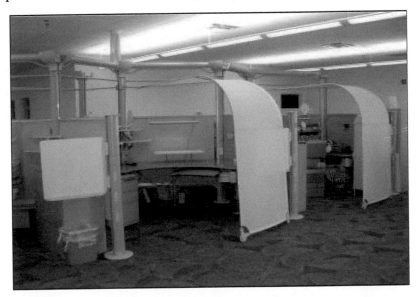

**Figure 6.34** Typical configuration of experimental/prototype space (Author))

- Acoustic and visual privacy;
- Adjustment;
- Availability of places to conduct individual work;
- Comfort;
- Confidentiality of materials and conversations;
- Distractions;
- Layout and size of workstation features;
- Natural and task lighting; and
- Storage.

Based on our findings, we show that there were statistically significant positive evaluations on five attributes, including: lighting, privacy, handling confidential materials, adjustability, and options for different places to conduct work (see Table 6.04). We also found there to be only one case in which the evaluation became more negative after the move to the new space, which was the evaluation of the amount of natural lighting available. See Table 6.04 for an overview of the Workspace Evaluation results for the Experimental group at the pre-move (Mean Group 1) and post-move (Mean Group 2) levels.

*Lighting.* Employees reported that the quality of workstation overhead lighting and task lighting had improved by 8%, compared to their assessments of lighting in their original workstations before their move to the new space (see Table 6.04).

Table 6.04 Workspace Evaluation: Experimental Group Pre/Post-Move

| Workspace (Attributes) Evaluation | | | | | | | | |
|---|---|---|---|---|---|---|---|---|
| Comparison of Pre-Move and Post-Move Data for Experimental Group | | | | | | | | |
| Question | N Group 1 | Mean Group 1 | N Group 2 | Mean Group 2 | DF | T Value | Prob | Statistically Significant Change |
| 1. I can comfortably use my workstation for long periods of time. | 19 | 3.95 | 34 | 3.88 | 51 | 0.43 | - | NA |
| 2. The arrangement of work surface, storage, chair, computer, etc. in my workstation supports my work needs. | 19 | 3.68 | 34 | 3.38 | 51 | 1.21 | - | NA |
| 3. The size of the work surfaces (desk, tables) in my workstation is appropriate for the work I do. | 19 | 3.21 | 34 | 3.06 | 51 | 0.42 | - | NA |
| 4. The lighting in my workstation (overhead, task) is appropriate for the work I do. | 19 | 3.58 | 34 | 3.91 | 51 | -1.46 | 0.1 | +8% |
| 5. The amount of natural light at my workstation is appropriate for the work I do. | 19 | 3.84 | 34 | 3.21 | 51 | 2.16 | 0.025 | -16% |
| 6. I am satisfied with the overall acoustical privacy provided by my workstation. | 19 | 2.21 | 34 | 1.97 | 51 | 0.77 | - | NA |
| 7. I am satisfied with the overall visual privacy provided by my workstation. | 19 | 2.37 | 34 | 3.18 | 51 | -2.81 | 0.005 | +25% |
| 8. My workstation supports my need to have face-to-face confidential conversations. | 19 | 1.68 | 34 | 1.65 | 51 | 0.13 | - | NA |
| 9. I feel comfortable working with confidential materials in my workstation. | 19 | 2.42 | 34 | 2.82 | 51 | -1.33 | 0.1 | +14% |
| 10. Other people distract me when I am working in my workspace. | 19 | 3.58 | 34 | 3.79 | 51 | -0.96 | - | NA |
| 11. I can adjust my workstation to fit my needs. | 19 | 2.63 | 34 | 3.35 | 51 | -2.89 | 0.005 | +21% |
| 12. I am satisfied with the availability of different workspace settings in which I can accomplish individual work. | 19 | 2.79 | 32 | 3.38 | 49 | -2.45 | 0.01 | +17% |
| 13. The amount of storage in my workstation meets my needs. | 18 | 3.22 | 34 | 3.26 | 50 | -0.13 | - | NA |

*Workspace Flexibility.* Employees reported that their satisfaction with the availability of different workspace settings in which to accomplish individual work, significantly improved by 17 percent compared to their assessments on this aspect of the workspace, before their move (see Table 6.04).

*Visual Privacy.* Employees reported that their satisfaction with overall visual privacy in their workstation significantly improved by 25 percent, compared to their assessments of privacy within their original workstations before their move (see Table 6.04).

*Working with Confidential Materials.* Employees reported that their comfort level in working with confidential materials in their workstation, significantly improved by 14 percent, compared to their assessments of privacy within their original workstations before their move (see Table 6.04).

**Figure 6.35** An enclosed workspace before the move (left), and a workspace in the new open work environment (right) (Author)

*Workstation Adjustability.* Employees reported that their ability to adjust the workstation components to fit their needs significantly improved by 21 percent, compared to their assessments of workstation adjustability within their original workstations before their move (Table 6.04). While not assessed in this study, previous research has found links between workstation adjustability and job control. Job control has been associated with long-term health effects and reduced risk of injury.

## 2. Workstation Evaluation Recommendations

In our findings, a key result reported was that employees experienced a significant increase in their ability to adjust their workstation to fit their work needs. Workstation adjustability has benefits related to better support for work process and healthful work.

We recommended that future workstation designs continue to incorporate enhanced adjustability of the workstation features. This may enhance the ease with which people accomplish their work. In other workplace research, workstation adjustability is linked to enhanced job control, and in the health research literature, enhanced job control is linked to reductions in health risks to employees. Thus, providing workstations with greater adjustability may enhance job control and reduce health risks to employees -- two powerful benefits.

## 3. Collaboration

Physical spaces to support meetings and collaboration were implemented near the professional workers' workstations in the Experimental group (see Figure 6.36 and right photo of Figure 6.37).

**Figure 6.36** A collaborative area within the experimental work environment (Author)

In the Collaboration section of the survey, we asked employees to evaluate their work environment on four attributes related to support for collaboration, including:

> - Design of meeting spaces;
> - Availability of meeting spaces;
> - Appropriate technology in meeting spaces; and
> - Workspace support for the quick shift from individual to collaborative work.

We found statistically significant positive evaluations on two of the four attributes of collaboration, and we found no cases in which evaluations became more negative after the move to the new space. See Table 6.05 for an overview of the workplace Collaboration results for the Experimental group at the pre-move (Mean Group 1) and post-move (Mean Group 2) levels.

Table 6.05 Collaboration: Experimental Group Pre/Post-Move

| | N Group 1 | Mean Group 1 | N Group 2 | Mean Group 2 | DF | T Value | Prob | Statistically Significant Change |
|---|---|---|---|---|---|---|---|---|
| **Collaboration** | | | | | | | | |
| Comparison of Pre-Move and Post-Move Data for Experimental Group | | | | | | | | |
| 1. The design of the various meeting spaces in this office provide adequate support for collaboration. | 19 | 2.63 | 34 | 3.5 | 51 | -3.29 | 0.001 | +25% |
| 2. Appropriate space is available for me to collaborate when I need it. | 19 | 2.74 | 34 | 3.35 | 51 | -2.27 | 0.025 | +18% |
| 3. Appropriate technology tools (display, video conference, capabilities, power/data connection, etc.) are available in the meeting spaces. | 19 | 3.58 | 33 | 3.7 | 50 | -0.58 | - | NA |
| 4. Overall, the workspace lets me quickly shift from individual work to collaboration with others. | 19 | 3 | 34 | 3.18 | 51 | -0.78 | - | NA |

***Design of Meeting Spaces.*** Employees reported that their satisfaction with design of various meeting spaces within the facility, significantly improved significantly by 25 percent, compared to their assessments of meeting space design before their move (see table 6.05).

***Availability of Meeting Spaces.*** Employees reported that their satisfaction with the availability of appropriate meeting spaces within the facility when needed, significantly improved by 18 percent, compared to their assessments of availability of appropriate meeting space before their move (see Table 6.05).

## 4. Collaboration Recommendations

In this study, we found consistent improvements in the design and availability of appropriate meeting spaces. We also found a decline in satisfaction with ability to have small meetings within the workstation.

We recommend that future facility designs make sure to incorporate an adequate variety of meeting spaces located near workstations, to facilitate having different types of small unplanned meetings – the type that had been held within workstations.

This shifting of group work (small group meetings) from the workstation to adjacent meeting spaces has several benefits. First meeting rooms can more easily be designed than workstations with the technology to support the needs of groups -- even for brief unplanned meetings. Second, moving from workstation to meeting rooms enhances potential for unplanned interactions with other employees, and provides a health benefit of getting employees to move around the space rather than spending too much time in static postures within the workstation.

## 5. Group Work Process

In the Group Work Process section of the survey, we asked employees to evaluate their work environment on six attributes related to quality of internal group process, including:

- Degree to which all members are involved in the activities of the group;
- Level of participation by members;
- Supportiveness;
- Participation in decision-making;
- Group has access to required technology tools; and
- Group has knowledge of how to use technology tools.

Table 6.06 Group Work Issues: Experimental Group Pre/Post-Move

| | | | | | | | | |
|---|---|---|---|---|---|---|---|---|
| **Group Work Issues** | | | | | | | | |
| Comparison of Pre-Move and Post-Move Data for Experimental Group | | | | | | | | |
| Question | N Group 1 | Mean Group 1 | N Group 2 | Mean Group 2 | DF | T Value | Prob | Statistically Significant Change |
| 1. All group members are involved in the activities of the group. | 17 | 3.35 | 32 | 3.09 | 47 | 1.09 | - | NA |
| 2. Participation by members is very high. | 17 | 3.47 | 32 | 3.44 | 47 | 0.16 | - | NA |
| 3. Group members are supportive toward each other. | 17 | 3.82 | 31 | 3.87 | 46 | -0.21 | - | NA |
| 4. Group members participate in decision-making. | 17 | 3.53 | 32 | 3.66 | 47 | -0.56 | - | NA |
| 5. My workgroup has access to the technology tools we need to effectively perform our work. | 17 | 3.94 | 33 | 3.73 | 48 | 1.29 | - | NA |
| 6. My workgroup possesses adequate knowledge of how to use technology tools in order to perform effectively. | 17 | 3.71 | 33 | 3.73 | 48 | -0.12 | - | NA |

We found no statistically significant changes (either positive or negative) on any of the six measures of Group Work Issues, after the move to the new work environment. See Table 6.06 for an overview of the Group Work Issues results for the Experimental group at the pre-move (Mean Group 1) and post-move (Mean Group 2) levels. These Group Work Issues were measured by assessing involvement and participation in the activities of teams. In our research, we rarely find significant changes on this measure. When we do find changes they are usually linked to ongoing organizational change or training programs that are affecting group work process and dynamics.

## 6. Communication

In the Communication section of the survey, we asked employees to evaluate their work environment on four attributes of communication, including:

-Face to face communication;
-Ability to access co-workers when needed;
-Accessibility to co-workers; and
-The ability to conduct small meetings in workspace without disturbing others.

We found a statistically significant negative evaluation on one of the four attributes of communication, reported by employees after the move to the new space. See Table 6.07 for an overview of the Communication results for the Experimental group at the pre-move (Mean Group 1) and post-move (Mean Group 2) levels.

Table 6.07 Communication: Experimental Group Pre/Post-Move Comparison

| Communication | | | | | | | | |
|---|---|---|---|---|---|---|---|---|
| Comparison of Pre-Move and Post-Move Data for Experimental Group | | | | | | | | |
| Question | N Group 1 | Mean Group 1 | N Group 2 | Mean Group 2 | DF | T Value | Prob | Statistically Significant Change |
| 1. I am satisfied with the overall quality of face-to-face communication I have with co-workers in this office. | 18 | 3.72 | 34 | 3.71 | 50 | 0.09 | - | NA |
| 2. It is easy for me to access co-workers when I need to discuss a work related issue. | 18 | 3.83 | 34 | 3.74 | 50 | 0.47 | - | NA |
| 3. I am easily accessible when others want to locate me. | 18 | 3.94 | 33 | 3.97 | 49 | -0.15 | - | NA |
| 4. I can conduct small group meetings in my workspace without disrupting my neighbors. | 18 | 4 | 34 | 1.76 | 50 | 9.76 | 0.001 | -56% |

Employees reported that satisfaction with their ability to hold small group meetings in their workspace without disturbing others, significantly decreased by 56 percent, compared to their assessments on this issue before their move (see Table 6.07). Clearly, the design of the new workspaces does not support the ability to have small group meetings within the workspace as well as the original workspace design. However, when this issue is considered in the context of the overall workplace design, employees reported a 25 percent increase in satisfaction with the design and availability of adjacent spaces in which to conduct meetings (see Table 6.07).

In addition, employees also reported an 18 percent increase in the overall design of various meeting spaces in the office that support collaboration (see

Table 6.07). Thus, this research indicates that small meetings have been shifted from within the workspace, to adjacent spaces.

In the Communicating Corporate Culture (Sense of Community) Through Design section of the survey, we asked employees to evaluate the degree to which their work environment communicates the image of the company to employees, enhances a sense of belonging to the organization, and the retention of talented people. This evaluation of the work environment was done on three attributes including:

- Design supports identification and feeling a part of the company;

- Space conveys appropriate image of company to employees; and

- Design of space helps retain talented people.

We found a statistically significant positive evaluation on one of the three attributes of "Communicating Corporate Culture" reported by employees after the move to the new space (see Table 6.08).

Table 6.08 Communicating Corporate Culture through Design: Experimental Group Pre/Post-Move Comparison

| Communicating Corporate Culture Through Design | | | | | | | |
|---|---|---|---|---|---|---|---|
| Comparison of Pre-Move and Post-Move Data for Experimental Group | | | | | | | |
| Question | N Group 1 | Mean Group 1 | N Group 2 | Mean Group 2 | DF | T Value | Prob | Statistically Significant Change |
| 1. The design of my workspace helps me to personally identify with, and feel I am part of, this company. | 18 | 3.33 | 34 | 3.21 | 50 | 0.86 | - | NA |
| 2. The office space in which my workspace is located conveys the appropriate image of this company to employees | 18 | 3.22 | 34 | 3.38 | 50 | -0.69 | - | NA |
| 3. The design of this office space helps us retain talented people. | 18 | 2.72 | 34 | 2.94 | 50 | -1.03 | 0.1 | +7% |

***Design of Space Helps Retain Talented Employees.*** Employees reported that their perception that the new workspace helps to retain talented people increased by 7 percent, compared to their assessment of the workspace on this issue before their move to the new space.

Thus, employees feel that the design of the new workspaces does impact the organizations' ability to retain talented employees, which is an outcome that is critical to the future success of the organization.

Research shows that part of the reason for this increased perception is that the new workspace creates a sense of professionalism in its' appearance, and a sense of belonging, through the overall design of space, and through features such as scanned imagery on workstation screens. We recommend that more screens be converted to display images, as we believe that this type of feature is an effective way to increase employees' sense of belonging and impact on employee retention.

**Figure 6.37** A work area before the move (left image) and after (right image) (Author)

### E. Conclusions

Overall, this study shows positive results for employees who moved from the original workspace, to the new workspace. Employees reported improvements in lighting, adjustability, ability to handle confidential materials, privacy, and flexibility to use alternative spaces to conduct individual work. The new workspace also supports collaboration more effectively through improved design and availability of spaces for meetings. We found no negative effects on the quality of group process for employees who received the new work environment. The only negative finding of the report was a significant decline in employees' ability to conduct small meetings in their workspace without disturbing others. However, this is offset by the redesigned space, which has improved the design and availability of alternative meeting spaces for employees. In terms of potential effects on retention, employees reported that they feel the new workspace helps retain talented employees more effectively than the original space they moved from.

## VIII. CASE STUDY: COLLABORATIVE SPACES
## FOR R&D JOBS

### A. Purpose of the Project

The project examined the effects of a more open work environment design concept using open workstations and collaborative spaces. The intention of this new environment is to enhance group collaboration and to build a sense of belonging to the organization for research and development (R&D) professionals.

### B. Work Environment and Participants

This project provided a comparison of employee behaviors and perceptions between employees within a mix of traditional private offices and solid panel-walled cubicles (providing seated, not standing height enclosure), and employees who moved into workstations that are, by comparison, very open, using a thin fabric "wall" to provide visual enclosure. In the new space, a variety of small group collaborative areas were also created and furnished to provide additional support for team work. Figure 6.38 shows an example of one of the casual meeting spaces.

**Figure 6.38** Open meeting space (Author)

Some of these spaces are enclosed and others are not. The idea is to have a variety of spaces that can be used to support unplanned collaboration between different numbers of people.

In the new space, the individual workstations themselves were organized into clusters or small groups (see Figure 6.39). This was a significant change from the rows of cubicles and enclosed private offices that employees moved from, which emphasized solitary work and did not support collaboration.

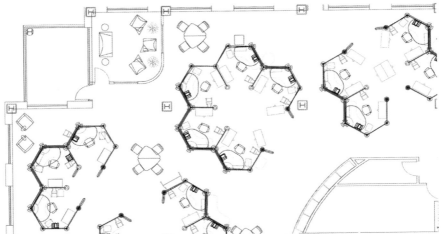

**Figure 6.39** Floorplan showing clusters of workstations (Author)

A total of 180 research and development (R&D) employees participated in the study, half of which served in the "experimental" group (the group that moved into a modified workspace).

## C. Study Hypotheses

The study explored several areas related to workplace collaboration, the quality of internal workgroup processes, and the sense of belonging to the organization by employees. We developed the following hypotheses:

- Collaboration will improve as a result of the effects of the more open work environment.
- Quality of internal group process will improve as a result of the effects of the more open work environment.
- Sense of employee identification and belonging to the organization will improve as a result of the new work environment.
- These effects will be found regardless of job type within the organization.

## D. Research Design

This study examined two groups of employees, about 180 participants, at two points in time. One group served as a Control group, (in which there were no changes made to their workspace) and the other group received the experimental intervention of a new workspace using open workstations and furniture. This type of study is known as a pretest-posttest design with a Control group.

Data was gathered from both the Control and Experimental Groups at two points in time, once before and once after the move to the new space. The data was collected using an electronic survey system. An independent statistician, who was not involved with the data collection of the study, conducted additional data correlations and factor analyses.

## E. Variables: Predicting the "Effectiveness of the Workplace"

Eight variables were used to assess the amount of satisfaction that the participants had with their workplaces. These were variables that indicated how much the participants thought certain elements of the workspace were effective in assisting them in their work (by supporting collaboration, group process, and feelings of belonging to the organization). These variables were combined to make up a single "Effectiveness of the Workplace" variable (all eight variables are listed in Table 6.09).

A technique called Factor Analysis was used to confirm that these eight variables group together and represent the construct of "Workplace Effectiveness." A factor analysis was performed on all variables used in the survey to uncover any patterns of response. From this analysis, we created an "Effectiveness of the Workplace" Factor (see Table 6.09). This analysis revealed that the respondents considered these items as a single group, and in fact all of these variables were significantly correlated with each other (in essence, participants who agreed with one variable, strongly tended to agree with all of the others).

As a result, this set of variables can be taken together as a single outcome variable, or factor, reflecting the overall satisfaction with the effectiveness of the workspace. A new "Effectiveness of the Workplace" variable was computed by adding scores of these variables.

## F. Analyses: Did the New Work Environment Make a Difference?

### 1. Data Set

We created a data set that included the responses of 143 cases to 25 survey variables, plus the variables of Job Type and Department. Not all of the cases had both pretest and posttest data, so the analyses were run with fewer cases than that of the entire file (roughly 90 cases for each analysis: 60 or so in the Experimental Group, and roughly 30 in the Control Group). This technique is called "matching" and it is the most sensitive and powerful analysis technique for this type of data.

| Table 6.09 Performance Variables: Effectiveness of the Workplace |
|---|
| The workspace helps team members feel like they are really part of the team through design features or visual cues. |
| The workspace supports team member participation in the ongoing work. |
| The design of the various spaces in this office provides adequate support for collaboration. |
| The overall workspace lets me quickly shift from individual work to collaboration with others. |
| It is easy for you to physically access co-workers when you need to discuss a work related issue. |
| The design of the interior space where my primary workstation is located contributes to my sense of belonging to the Sara Lee organization. |
| This office space conveys the appropriate image of the company to employees and others. |

### 2. Outcome Factor

We then used the new "Effectiveness of the Workspace" factor as an outcome factor in the two main analyses for this study. An outcome factor is a factor that changes as a result of some change that was made to the work environment. For example, "privacy" could be an outcome factor. We might decide to put up higher office walls and test to see if the outcome factor of privacy will increase. Thus, in this study we are testing to determine the effects of the new workspace on the outcome factor of *"Workplace Effectiveness."*

### 3. General Analyses

In the first analysis, we looked for differences in the "Effectiveness of the Workplace" between the Experimental and Control groups. We were seeking to find whether the new work environment design strategy had any impact on employees' assessment of the "Effectiveness of the Workplace." In the second analysis, we determined which workplace design characteristics were the most important predictors of the "Effectiveness of the Workplace" for each functional job category (Job Type).

*Analysis Between Experimental and Control Groups on the "Effectiveness of the Workplace."* We calculated a t-test, which showed that the Experimental group had a significantly higher overall satisfaction with "Effectiveness of the Workplace," than did the Control group (t (88) = 1.99, $p$ <.025), and this advantage held for all categories of Job Type.

Thus, based on these findings we can state that the new work environment has been shown to provide a significantly better (more effective) work environment than the existing workspaces (used in the Control Group) regardless of the type of work performed. Note that this result is based on employees' own assessment on what constitutes an effective work environment (support for group process, collaboration, and belonging to the organization).

4. Detailed Analysis

There were significant differences between the Experimental and Control groups with respect to the change in the eight "workplace" variables. In each case, the change was positive for the Experimental group, and negative (or no change) for the Control Group. Table 6.10 lists each of the survey question items for which we found significant differences between the means.

5. Interpreting the Results

The left column of Table 6.10 shows the survey question item for which there was a significant difference. The middle and right columns of Table 6.10 show the amount of change between the pre-move and post-move survey scores for each question item (based on a 5-point Likert-type scale). Consider the first item, "Satisfaction with being overheard while talking in a normal voice in your workstation." The change to the Experimental Group from before they moved, to after moving to the new workspace is +.1, indicating a slight increase in satisfaction with this workspace characteristic.

The change to the Control group, which received no change to the work environment, but was simply measured at the same times as the Experimental group, was negative, -.23. The statistical analysis then showed that when compared to the amount and direction of change to the Control Group, the slight positive change to the Experimental group is statistically significant, meaning that employees in the new Resolve workspace are more satisfied with the degree to which they are overheard while talking in a normal voice in their workstation.

This result may seem counterintuitive since an open workstation environment is a more open workspace than the traditional workspace that the employees moved from. However, employees working in open environments are more visible and they can give and receive non-verbal cues from fellow employees about distance and the potential to be heard, which will allow people to modulate their voices to prevent being overheard. This is a behavior that is not possible in a more physically enclosed (but just as acoustically open) workspace.

Table 6.10 Summary of Differences between Control and Experimental Groups on Workspace Characteristics

| Workspace Characteristic | Change to Experimental Group | Change to Control Group |
|---|---|---|
| Satisfaction with being overheard while talking in a normal voice in your workstation. | +.1 (+2.97%) | -.23 (-6.01%) |
| Ability to handle confidential materials (papers on my desk, computer screen). | +.37 (+13.70%) | -.29 (-9.80%) |
| Distractions from background noise (air conditioning, printers, fax machines, etc.). | +.16 (+6.15%) | -.29 (-6.18%) |
| Overall sense of privacy. | +.36 (+11.84%) | -.23 (-6.63%) |
| The ease with which you can organize work materials in your workstation/office. | +.57 (+26.39%) | .00 (0%) |
| Frequency of scheduled meetings. | +.10 (+3.53%) | -.23 (-6.71%) |
| Conference room elsewhere in the building. | +.41 (+13.71%) | -.10 (-3.89%) |
| The design of the interior space where my primary workstation is located, contributes to my sense of belonging to the Sara Lee organization. | +.03 (+.89%) | -.43 (-14.43%) |

## G. Summary of the Analyses Between the Experimental and Control Groups

In summary, the treatment (new open work environment) appears to have consistently improved conditions with respect to privacy, distractions, and the ease of organizing materials, as well as contributing to a sense of belonging.

In addition, the participants in the treatment condition (Experimental Group) spent more of their time in scheduled meetings (as opposed to unplanned one-on-one meetings) and they tend to have meetings in a conference room elsewhere in the building.

1. Analysis by Job Type

We analyzed the data to identify which workplace design characteristics are the predictors of the "Effectiveness of the Workspace" for each job type.

***Executives***
- Overall sense of privacy.
- Amount of distractions from over-hearing conversations of immediate neighbors.

***Directors***
- Amount of distractions from over-hearing conversations of immediate neighbors.
- Ability to have face-to-face confidential conversations.
- Visual distractions.

***Managers***
- Amount of distractions from over-hearing conversations of immediate neighbors.
- Ability to have face-to-face confidential conversations.

***Professional/Technical***
- Arrangement of furnishings and equipment (work surface, storage, chair, computer, etc.) within the workstation.
- Availability of technology/tools required to do the job.
- The ease of organizing work materials in the workstation/office.
- Distractions from background noise (air conditioning, printers, fax machines, etc.
- Visual distractions.
- Ability to have face-to-face confidential conversations.

Notice the shift from basic privacy concerns (Executives) to the ability to have confidential conversations (Directors, Managers), to the basics of the physical layout of the workstation (Professional/Technical). This study has revealed some basic differences in what people in various job categories consider to be important to their ability to work in a physical workspace. This analysis can be used to inform programmatic requirements by job type for future implementation of the space.

## H. Conclusions

The study hypotheses were as follows:

1. Collaboration will improve as a result of the effects of the more open work environment.

2. Quality of internal group process will improve as a result of the effects of the more open work environment.

3. Sense of employee identification and belonging to the organization will improve as a result of the new work environment.

These effects will be found regardless of job type within the organization.

In all cases, the results of this research show that these hypotheses were supported.

Regardless of job type or department, all participants experienced positive, beneficial effects of the new workspace using Resolve on key project goals, which included: improved collaboration, enhanced quality of group process, and enhanced sense of belonging to the organization.

# CHAPTER 7

## Case Studies: Work Space and the Individual

### I. EXECUTIVE SUMMARY

The potential for environmental control at the individual level is determined by the adjustability of the physical elements within the workstation, such as: task lighting, shelves, storage, work surface height, level of enclosure, VDT screen and keyboard, and HVAC system (Heating, Ventilation and Air Conditioning). Control over these elements may contribute to comfort, environmental satisfaction, privacy, communication, and other perceptions that are related to effectiveness and quality of work life. Many of these features are also related to ergonomic and health considerations. At the individual level, research suggests that environmental control over workstation components has a direct relationship to performance. A flexible work environment may also enhance job control, by permitting adjustments to the physical environment to be made according to individual task demands or the needs of team workers at a particular stage of a project.

This chapter describes five case studies that focus on user control over the work environment. Unless otherwise noted these studies were conducted by the author. The relation between control and outcome measures, including stress and work performance, is examined. We have portrayed a variety of job types, including: Project Managers, Sales Professionals, Administrative staff, Executives, Call Center workers, Managers, Accountants, and Real Estate Management professionals. This work occurred in many types of work environments. The range of studies helps us to generalize our conclusions to as wide a range of workers and environments as possible. We believe that even with the relatively small number of case studies presented here, the basis for a sound argument exists for the use of environmental control concepts as a mechanism for enhancing work environments.

*Case Study 1.* The first Case Study describes a project in which we worked with a high technology manufacturer who was concerned about the effects of work environment design on the health of employees in their Field Sales offices. The client was especially interested in the potential of user control over the environment as a tool to enhance worker health. The company has field sales offices located in every major metropolitan area of the U.S. The purpose of the project was to identify the work environment features that, if not present or poorly designed, would constitute a health risk (psychological stress, musculoskeletal pain, and general health) to those workers. The deliverable for this project was the translation of this information into recommendations for workstation design and layout, based on the results of this project.

*Case Study 2.* The second Case Study describes a project in which we worked with a large company in the retail clothing business, and examined the relationships between ambient environmental conditions and work space design features, and reactions of workers in two Call Centers. The work environment variables we studied included: air quality, noise, work space layout, and adjustability of interior work station components. The employee reactions we studied included: stress and musculoskeletal complaints, job satisfaction, and self-assessed performance. The analysis led to conclusions about the relations between certain environmental features in these work environments, and worker reactions, particularly performance and stress.

*Case Study 3.* The third Case Study examined the usefulness of a workstation design concept, "active storage," in supporting computer-intensive jobs in which paper-based tasks need to be integrated. We believed that active storage enhances user control over the environment, and thus the potential to enhance health and performance. We examined the effectiveness of this design concept with two different job types: Senior Project Managers (building design and construction responsibilities), and Accountants. These participants worked for two different companies in different industries, in different areas of North America. While the functions and responsibilities of these job types varied dramatically from each other, they have in common work processes that included intensive paper-based tasks as well. We evaluated the usefulness of the design concept in terms of how it might enhance health and work effectiveness. Reactions to working in the experimental environment were investigated, including: satisfaction with workspace, visual and auditory privacy, psychological stress, musculoskeletal problems, group effectiveness, and individual performance. Other perceptions of support for the work process were also examined, including: the degree to which the workspace aids efficiency, quality of storage, flexibility of workspace, and control over pace of work. The deliverable of this project was understanding about the contribution of this design concept to supporting the work process of these workers, and specific links of this design to health and performance.

*Case Study 4.* The fourth Case Study describes a project in which we worked with a high technology manufacturer who was concerned about the effects of work environment design on the performance of individual workers and effectiveness of teams in their Field Sales offices. The client was especially interested in the potential of user control over the environment as a tool to enhance worker performance. The purpose of the project was to identify the work environment features, particularly those related to control over the physical work environment, that enhance individual and group performance. The deliverable for this project was the translation of this information into general recommendations for workstation design and layout.

*Case Study 5.* The purpose of this laboratory experiment was to examine the effects of interior workstation adjustability - and the effects of training in how to use that adjustability - on physiological stress and motivational performance levels under high workload. We hypothesized that more control (through adjustability) and training would have the most desirable impact on these outcomes.

Statistically significant findings of the study suggest:

- Given the opportunity through workstation adjustability, people will exert control over the work environment.
- Physiological signs of stress are reduced when people have workstation adjustability and the training to use it.
- Motivational performance is enhanced when people have workstation adjustability and the training to use it.

Taken in concert with prior field work on adjustability in the workplace, the results from this laboratory study lend some support to the claim that control in the form of adjustability of workplace features is an important element in stress and performance motivation. The findings also underscore the critical role of proper training. Provision of office furniture and technology that increase the range of user adjustability will, in general, not lead to benefits unless the worker is instructed in the proper use of increased opportunities for adjustability.

## II. CASE STUDY:
## WORKSTATION FEATURES THAT PROMOTE
## HEALTHY WORK FOR DIFFERENT JOB TYPES

### A. The Project

We conducted a project with a high technology manufacturer who was concerned about the effects of work environment design on the health of employees in their Field Sales offices. The company has field sales offices located in every major metropolitan area of the U.S. The purpose of the project was to identify the workstation design features and capabilities that could promote healthy work in those offices. Looked at another way, we wanted to identify the work environment features that, if not present or if poorly designed, would constitute a health risk (psychological stress, musculoskeletal pain, and general health) to those workers. In addition, we recommended to the client that we develop this information for each type of job within the Field Sales offices. We believe that because jobs differ in their demands and work process characteristics, workstation features related to health for one job type might be less important to healthy work for another job type. The deliverable for this project was information detailing the research findings, as well the translation of this information into recommendations for workstation design and layout, based on the research results.

### 1.   Work Environments

The office environments in use for the sites that we included in our sample ranged from traditional dry wall, fully-enclosed private offices, to open plan systems furniture enclosed by panels. Figure 7.1 shows a typical example of the layout of a facility using private offices.

Figure 7.2 shows a photograph of the interior of a typical private office in use at these locations. These offices used primarily steel and wood "casegoods," having freestanding desks and storage for books and paper files. Most of the private offices had windows to exterior views.

### 2.   Job Design

Jobs, like environments, have design features. The design features of jobs are not as readily apparent as the design of work environments. A job design may include characteristics such as pace of work, amount of control over the job, autonomy, work goals, team membership, and a variety of other facets.

**Figure 7.1** Typical layout of facility using private offices (Author)

Thus, the design of a job means that it will have specific characteristics that will be different from other jobs.

Different jobs have distinct "profiles" of demand characteristics. Demand characteristics have to do with the psychological demands of a job - the intensity of the mental work of a job. Demands are measured through workload, time pressure, and responsibilities, all of which are related to psychological stress (Carayon, 1994). The cumulative effects of job stress over time can lead to serious health consequences, such as hypertension and coronary heart disease (Karasek and Theorell, 1990).

## 3. Job Types

We identified four distinct types of jobs that exist within the Field Sales function of this company, including: Sales, Customer Technical Support, People Managers, and Administrative Staff.

The sales jobs are highly skilled positions in which management of a sales process occurs. In the sales process, technical and analytic resources are managed in the process of analyzing and solving customers' problems as well as the traditional process of creating a relationship with the customer and closing a sale.

**Figure 7.2** Typical interior view of private office (Author)

This job functions primarily in a leadership position within a cross-functional team, coordinating subprocesses around the process of identifying and solving customers' technology problems. These teams are small and the work is ongoing. Several projects may be in various stages of completion at a given time. In our analysis, the characteristics of this job reminded us of the process of creating and managing long term customer engagements that occurs in consulting companies.

The role of the technical support job is to provide expertise regarding software and hardware for the ongoing support of the customer's computer system. This job type works as an individual, part of a larger group of other technical support workers. Despite the individual nature of the work, our interviews with workers having this job revealed strong group cohesion and identity. The job includes being responsible for compatibility issues with competitive hardware and software, installation of software and hardware upgrades, and communicating information to the client's computer support staff. Upon purchase of a computer system, the client is assigned a permanent technical support person who is then given a permanent office at the client's site. A worker with this job will support more than one client at a time. Computer technical support professionals spend the majority of their time at their office at the client site(s), providing technical support. When not at the client site, each technical support worker has a non-dedicated office space within the field sales office.

The role of people managers within Field Sales is to direct the overall sales strategy for the local unit, and lead a functional team of sales professionals. Each Field Sales unit may have several managers, each of whom has a number of sales professionals reporting to them. These managers have formal authority over the strategic sales planning process and the personnel reporting to them.

The role of administrative staff is to provide a support function for the activities of the sales teams and managers. To a lesser extent they provide indirect support for technical support professionals. Administrative tasks include telephone work, reception, typing and managing proposal document development, and other tasks typical of administrative work. This work style is individual, not team based.

## B. Methods

### 1. Hypotheses

We hypothesized that workstation design characteristics act as risk factors to employee health, but that the design characteristics that might affect health will differ depending on job type. Thus, different workstation design features will act as stressors for different jobs. Previous research shows that different jobs have different demands and work processes, and thus design features that present health risk factors to one job type may not represent risks to another job type (Carayon, 1994). Our intent was to understand which design features are related to health for particular job types, so that we could make design recommendations to create healthy environments for each job type.

### 2. Data Collection and Analysis

A confidential questionnaire was used to collect data from 66 workers in four job types, located in four field sales offices, in two states. Job categories (with number of respondents per category in parentheses) in the sample included: professional sales people (18), computer technical analysts (34), people managers (8), and administrative staff (6).

To test the hypothesis, regression models with workstation design features as predictors, including: lighting, auditory and visual privacy, support for status, ability to personalize, adjustability of workstation design features (control), fit of workstation layout to the job (layout), and quality of storage (storage), were computed for each of the four job types. Indices of employee health were created to serve as outcome measures, including: self-reported stress, musculoskeletal pain, and general health. Other research suggests that gender, age, and amount of time spent in the workstation might influence the stress and health of workers. To remove the effects of these variables, they were entered as a block in each regression equation during our analysis. This was done so we could determine the true effects of control, layout, and storage on employee health.

## 3. Measures

**Design Features and Capabilities.** Multiple-item measures of the environmental control, layout and storage constructs were created. The reliability of these measures was tested using Chronbach's alpha ($\alpha$), a statistic that allows researchers to ask a number of questions relating to the same concept (for example, privacy) and then statistically collapse the responses to these questions into a single index representing that construct. Chronbach's alpha can range from 0 (representing a completely unreliable index) to 1 (meaning a highly reliable and useful index). The greater the alpha value, the more reliable the index.

Environmental control was measured through a six-item index of variables measuring aspects of interior workstation adjustability. This index provides information about the amount of control employees have over their immediate work environment. This six-item index assesses the adjustability and flexibility of chair, furnishings, lighting, and work materials ($\alpha = .76$). This index uses a 5-point Likert type scale where 1 = very dissatisfied and 5 = very satisfied.

Fit of workstation layout to job needs was measured through a seven-item index, including: suitable size of workstation floor area, convenient arrangement of furnishings, workstation support of efficient work, paper based materials close at hand, correct furniture for work performed, appropriate arrangement of furnishings and equipment, and work surface size appropriate for job ($\alpha = .88$).

Quality of storage was measured through two items: satisfaction with overall amount of storage, and satisfaction with the number of file drawers in the workstation.

**Health Outcomes.** Stress was measured through an eight-item index assessing short-term psychological stress reactions in workers during the past month, including: heart rate, breathing, appetite, dizziness, stomach problems, and trembling hands ($\alpha = .74$). This index used a 3-point Likert-type scale where 1 = "never" and 3 = "three or more times" that the respondent experienced a reaction in the last month (Caplan et al., 1975).

Musculoskeletal pain was assessed through measures of back pain, muscle/joint pain, upper body pain, twisting discomfort, and discomfort in hands. The scales for these measures used a 5-point Likert type scale where 1 = "never" and 5 = "always" experience pain in the particular body part (Sainfort, 1990).

## 4. Results

***How to Read the Results.*** Table 7.1 is an example table to help you understand how to interpret the findings, which are displayed in Table 7.2. In both tables, each job type is shown as a column head (see "Sales Jobs" at top of column in Table 7.1). The health outcome measures are listed to the left of the table as row headings (see "General Health," Table 7.1). For each set of "Job type" and "Health outcome" headings there is a corresponding cell within the table.

Within each cell there are two pieces of information. The first is a listing of the workstation features that are statistically significant predictors of that measure of worker health. We call these workstation features "risk factors" to health. Within each cell, some combination of workstation adjustability (control), fit of workstation layout to job (layout), or quality of storage (storage) are listed if found to be statistically related to that health outcome. Some cells have all three risk factors, and many have fewer or none at all (refer to Table 7.2). In our example table (see Table 7.1) "Layout" and "Control" are the design characteristics that may place a Sales worker's "General Health" at risk if not well designed. In addition, the workstation design risk factors in each cell are listed in descending order of their relative importance to the health outcome. Thus, in Table 7.1, "Control" is relatively more important than is Layout as it affects general health. This is only a relative difference, and taken together both features are important.

Table 7.1
How to Read the Results

|  | **Sales Jobs** |
|---|---|
| **General Health** | Control<br>Layout<br>$R^2 = .23$ |

The second piece of information is the "$R^2 = .xx$" number in each cell under the risk factors. This number is known as "R-Squared," and it can vary from zero to .99. Technically, the $R^2$ value is known as percentage of variance in the outcome measure (in this case, health outcomes) that is predicted by the risk factors in the cell. Thus, a value of .99 would mean that the risk factors in the cell almost perfectly predict the health outcome for that row in the table.

In more general terms, $R^2$ is an indicator of the predictive strength of the risk factors (in the case of our example, "Control" and "Layout") as they predict an outcome like "General Health" (see Table 7.1). Obviously many factors in the work and home environment predict an outcome such as "General Health," including: diet, sleep, alcohol intake, relationships with co-workers, marital status, and many others aside from the physical environment. Referring to Table 7.1 as an example, we find that $R^2 = .23$. This means that about 23 percent of the "general health" score is predicted by the environmental design factors of control and layout. This is a significant percentage and suggests that these design variables play an important role in the general health of sales workers.

***Results for Sales Workers.*** This case study shows that workstation adjustability (control), fit of layout to work process, and quality of storage, are workstation characteristics that have a great potential to affect the health of sales worker's - especially if those workers have a job design that is similar to those described in this case study. If these features or characteristics are designed poorly, they will constitute risk factors to sales workers health. For sales jobs, layout is a design characteristic that predicted problems with every aspect of worker health, from psychological stress to back pain and pain in wrist, fingers, and hand (muscle pain) (see Table 7.2).

Especially troubling is the strong relationship between poor layout and pain in hands and upper body for sales workers (see Table 7.2). Pain indications in hand, wrist, and fingers can be early indicators of Carpal Tunnel Disease (CTD). A single CTD case can cost an employer up to $80,000 including surgery, lost work time, and job re-training (Dainoff, 1982). However, many organizations report increased productivity after implementing ergonomic programs.

Thus, for workers with the type of job profile that the sales workers in this study had, poor design in terms of adjustability of the components within the workspace (control), fit of the layout of the workstation to job needs (layout), and quality of storage for paper work materials could pose significant health risks. Conversely, these findings show that there is great opportunity in focusing design and work environment resources around the issues of control, layout, and storage to create healthy work environments for sales workers (see Table 7.2).

***Results for Technical Professionals.*** For Technical Professionals, control and layout were weak but significant predictors of stress and general health (see Table 7.2). The nature of the technical professionals' job makes them less sensitive to design influences, since they spend much time at the clients' location.

**Results for Managers.** For Managers, none of the design features or capabilities predicted health or pain, but layout was a predictor of stress for these workers (see Table 7.2).

| Table 7.2 Workstation Risk Factors by Job Type and Health Problems | | | | |
|---|---|---|---|---|
| | **Sales** | **Technical Profes-sionals** | **Manag-ers** | **Administra-tive** |
| **General Health** | Control Layout $R^2 = .23$ | Layout Control $R^2 = .09$ | | Control $R^2 = .50$ |
| **Stress** | Control Layout Storage $R^2 = .19$ | Control Layout $R^2 = .05$ | Layout $R^2 = .69$ | Layout $R^2 = .96$ |
| **Back Pain** | Layout $R^2 = .38$ | | | |
| **Muscle/ Joint Pain** | Layout Storage $R^2 = .45$ | | | Layout $R^2 = .56$ |
| **Upper Body Pain** | Layout Storage $R^2 = .56$ | | | |
| **Twisting Discomfort** | Control Storage Layout $R^2 = .55$ | | | |
| **Hand Pain** | Layout Control $R^2 = .55$ | | | |

In fact, fit of workstation layout to the job predicted almost 70 percent of psychological stress for managers. This suggests an opportunity to carefully design workstation layout to match the needs of managers. The investment in time and resources to create a superior workstation layout could result in some stress reductions for these workers.

**Results for Administrative Workers.** For Administrative workers, layout is a very strong predictor of stress and muscle/joint pain (see Table 7.3).

The findings of this study suggest that appropriate layout could reduce psychological stress and muscle pain, both of which may be related to more serious musculoskeletal injuries. Thus for these jobs, we suggest an investment of resources to create appropriate workstation layout.

Environmental Control is a strong, significant predictor of general health for these workers, predicting up to 50 percent of the variance in general health scores. Thus, designing the workstation to enhance user control could result in significant benefits to workers with this job type, in terms of reducing their risk exposure to negative health consequences.

*Conclusions.* We believe that different jobs, with their related demands and processes, require different types of support through workstation design features.   We also believe that with the correct understanding, work environments can be purposefully designed to enhance the work process and reduce risks to worker health.  This is a significantly different approach to design development than traditional programming methods that soley address work process and organizational status issues.

Most workstation design is influenced by external issues such as job rank and corporate facility programs.  Because of the lack of understanding about the relationship between workstation design features, job design, and health risks, workstation design criteria cannot currently be developed to address healthy work.  Increasing our understanding of workstation features that can be manipulated by design could help office planners engage in the design process in a proactive manner, directing resources to minimize health risks to individuals, and ultimately, costs to the organization.

## 5.   Design Implications

In this section we discuss specific design features that should be considered when initially designing a workstation, or to be focused on when evaluating an existing workstation standard for possible re-design.  These recommendations are not tied to specific job types. To make specific linkages to job types, refer to Table 7.3.

Table 7.4 shows a summary of the health effects of the three groups of design characteristics we examined (control, layout, and storage) on each of the 4 job types in this case study. This table shows at a glance the health problems resulting from lack of workstation adjustability (control) on administrative jobs -- it affects perceptions of "general health" (see Table 7.4).

In this section, we discuss design recommendations for each of the three design features or characteristics that we examined in this case study, including control, layout, and storage (see Table 7.4).

*Environmental Control.* This characteristic is related to the amount of control the worker has over his immediate work environment. Aspects of control include adjustability of chair, furnishings, lighting, and work materials.

Select seating products designed to provide maximum control over adjustability by the user.

If you select seating that requires a high degree of active control by the user, make sure that users have adequate knowledge of how to adjust the seating. Knowledge of seating adjustment can be provided through in-house training programs or through outside vendors. The importance of implementation of training programs cannot be understated. We have seen many instances of workers seated in awkward postures, in some cases feet not touching the floor, because they were not aware that the expensive chair they were seated on could be adjusted.

Some furnishings within workstations can be adjusted by the user or fairly easily through the facilities group. These include shelving heights and in some cases position, hanging file tools, keyboard tray or surface, the primary work surface itself, filing cabinets on wheels, etc. Select these furnishings with adjustability/moveability as a primary criterion. Remember that it is important to train end users on the availability of the adjustments. Maximize the investment in adjustability through training programs. Work with managers to ensure that the work group culture supports the notion that items within the workstation can and should be moved or adjusted as conditions require. The management, and facility management style should encourage adjustment and movement of items within the workspace. While there is always the risk that highly moveable items will "walk off" to other areas, facility management problems such as that should not override the great advantage to worker health that adjustable components can provide.

Select task lighting with ease of movement and adjustability as a primary criterion. The worker should be able to adjust the light to eliminate glare, and coordinate the position of the light with the combination of computer and paper-based tasks that the job may require.

***Fit of Layout to Work Process.*** In this study, layout was measured through suitable size of workstation floor area, convenient arrangement of furnishings, workstation support of efficient work, paper-based materials close at hand, correct furniture for work performed, appropriate arrangement of furnishings and equipment, and work surface size appropriate for job. These are all characteristics of design that are related to proper fit of workstation to the work process.

***Adequate Floor Space.*** Does the workstation have adequate floor space to support the work that occurs within it? Is there adequate knee clearance under the work surface? Is there enough space for the seated worker to move about and access storage and different areas of the work surface easily?

| Table 7.3 Health Effects by Job Type and Workstation Features | | | | |
|---|---|---|---|---|
| | **Sales** | **Technical Professionals** | **Managers** | **Admin-istrative** |
| **Control** | General Health Stress Hand Pain Twisting | General Health Stress | | General Health |
| **Layout** | Stress Back Pain Muscle Pain | General Health Stress | Stress | Stress Muscle Pain |
| **Storage** | Stress Muscle Pain Upper Body Twisting | | | |

If frequent meetings occur within the workstation, is there adequate room to accommodate the visitor(s) and sufficient work surface to support collaboration? Is the workstation large enough (in terms of square footage) to support general movement related to work tasks and accommodate storage needs? (We have often observed what appeared to be reasonably sized workstations with the entire floor surface covered with piles of printouts, boxes, etc.)

*Arrangement of Furnishings and Worktools.* After observing the range of tasks and interactions occurring within the workstation, is the technology (computer, monitor, telephone, etc.) arranged in a way that minimizes awkward positions and appears to support the work tasks? (Think of situations in which the worker is on the phone, cord stretched to its limit, while on the computer or accessing paper files.) If the task demands require impossible arrangements of worktools, is the workstation or technology support designed so the user may easily relocate items on a short-term basis to support ongoing work?

*Support for work with paper-based materials.* Given the specific types of paper-based materials that the worker uses (binders, books, paper in folders, legal size paper or unusual size sheets, rolls of paper, etc.), are the workstation interior furnishings and support tools adequate for the job tasks? Watch over time as the work shifts from individual-based tasks to small meetings within the workstation. Do the various design elements within the workstation support these transitions in paper handling smoothly?

*Adequate work surface size.* Closely related to work process and paper based needs is the size and configuration of the work surface. This is especially important when the worker is combining paper and computer based work tasks, since the work surface must be designed to support paper display and VDT work. Consider the size, amount and type of paper, and other work materials used in different job tasks. Watch what happens when small meetings occur within the workstation. Is the work surface correctly designed or do materials end up scattered on file drawer tops, or laid out on the floor (causing awkward postures or poor viewing). The work surface size issue is closely related to overall square footage of the workstation, but creative design can lead to appropriate work surface even in constrained situations.

*Storage.* Quality of storage was measured through two items: satisfaction with overall amount of storage, and satisfaction with the number of file drawers in the workstation.

In your analysis of work needs, follow common sense in determining storage requirements. Rather than maximizing storage within the workstation at the expense of floor area and meeting room, provide adequate storage for "hot" documents and plan for longer term storage capability adjacent to the workstation. Small clusters of storage units that serve smaller groups of workstations provide ease of access to needed files without creating long distances to walk, or problems in locating the correct file drawer. It is important that workers maintain a sense of control over their paper files and that access to their files not disturb their work flow.

## III. CASE STUDY: CALL CENTER OPERATORS
## AND WORKSTATION DESIGN

### A. The Project

This study examined the relationships between ambient environmental conditions and work space design features, and worker reactions. The work environment variables we studied included: air quality, noise, work space layout, and adjustability of interior work station components. The employee reactions we studied included: stress and musculoskeletal complaints, job satisfaction, and self-assessed performance.

### B. Methods

1.  Settings and Participants

The settings were two Call Center facilities of a large mail order catalog company. The work done at these locations consists of taking orders for merchandise over the phone and processing the orders at a computer terminal. The operators also have responsibility for handling other customer requests for exchanges, refunds, and general information. These jobs have high psychological demands for accuracy and speed but low control over how to do the job or control over the pace of the work. This type of work has a design that can lead to high levels of psychological stress (Karasek and Theorell, 1990). The two sites use identical work spaces, except the panel heights for operators at one site are approximately 12 inches taller than the panels at the other site. The average square footage of work spaces was identical at both sites.

Figure 7.3 shows a drawing of a typical workstation for this job. The workstation includes: 1) a 3" drawer, 2) an articulating keyboard surface, 3) a 30" x 48" work surface, 4) a worktool for holding binders and reference materials, 5) paper management trays, 6) the mounting bar for the paper management tools, and 7) an 18" x 60" tackable surface.

Figure 7.4 is a photograph showing typical posture of an operator seated at a workstation. The background of this picture shows that from a standing height it is possible to see across the length of the entire facility and into many of the adjacent workstations. While the interior square footage within the workstation is somewhat constrained, the design of aisle widths permits operators to push back slightly from the work surface and "borrow" aisle space (see Figure 7.4). Operators also pushed back in this manner to briefly converse with each other between calls, while remaining in a seated position.

Figure 7.5 displays an additional view of two adjacent, unoccupied workstations. This photograph shows the location and placement of equipment and worktools within the space. Note the operators' headset hanging on the shared panel dividing the two workspaces. Operators also place paper materials on the interior surfaces of the panels.

***Demographics.*** A total of 385 employees answered questionnaires. 90 percent were female, 10 percent of participants were supervisors. The participants had an average of five years of computer experience and worked an average of 26 hours per week. Table 7.4 summarizes the demographic information for the two groups.

Of the 385 customer service representatives who completed the questionnaires, 99 worked at the site "A" and 286 at site "B" (see Table 7.4). Most respondents were females (89 percent), worked part-time (79 percent), were permanent employees (69 percent), and were in non-supervisory positions (89 percent). On average, they worked about 24 hours per week and had been employed with the company for about 3.5 years. Workers at site B had been employed with the company longer, worked more hours per week, and a greater percentage of respondents were supervisors than those at site A.

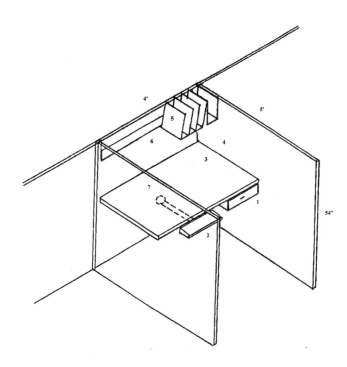

**Figure 7.3** Sketch of typical agent workstation layout (Author)

**Figure 7.4** Photograph of agent seated at typical workstation (Author)

2.   Outcome Variables

***Stress and Musculoskeletal Problems.*** Eight items were used to assess psychosomatic symptoms of stress, including: heart beating hard, shortness of breath, upset stomach or stomach ache, hands trembling enough to bother you, ill health affecting your work, loss of appetite, heart beating faster than normal, and spells of dizziness.

Musculoskeletal complaints (pain in arms/hands, neck and shoulders) and visual discomfort (eye strain) were also measured. These problems are thought to be health reactions to a stressful work environment.

***Job Satisfaction.*** Job satisfaction was assessed through a four-item scale measuring pride in company, sense of belonging to organization, and overall satisfaction with the job.

***Performance.*** Individual performance was assessed through self-evaluations on a nine-item scale measuring effort, quality and quantity of work, and errors. The items in this scale were based in part on a scale developed by Carayon (1994).

***Environmental Control.*** Two types of environmental control were measured: 1) control over the environment in terms of flexibility and adaptability of the work space, and 2) degree of exercised control over visual access.

Control as a function of flexibility/adaptability was assessed through two scales. The first is a three-item scale measuring adjustability and flexibility of computer, lighting, and furnishings in the work space.

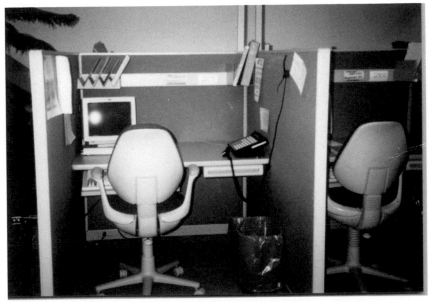

**Figure 7.5** View of typical operators' workstations (Author)

The second scale measured frequency of exercised control over air quality including: adjusting ventilation, opening/closing a window, adjusting temperature, and putting on/removing clothing to adjust temperature comfort. Control over visual exposure was measured on a four-item scale that assessed the frequency that employees exercised control over being heard, being seen by others, and controlling the number of people passing by or entering the work space.

3. Environmental Variables

***Work Space Support of Work*** was evaluated through a five-item scale measuring problems with amount of working surface space, arrangement of furniture and equipment, storage, work space appearance, and having to stretch or twist to reach things.

*Ambient Conditions.* Existing conditions within the office environment, including lighting, noise, temperature, humidity, air odors, and movement, were measured on multiple item scales.

| Table 7.4 Participant Demographics | |
|---|---|
| **Number of Participants at Each Site** | Site A n = 286 (74.3 %)<br>Site B n = 99  (25.7 %t) |
| **Gender** | Male n = 47     (7.5 %)<br>Female n = 338   (87.5 %) |
| **Job Type**<br>Part-time n = 302 (77.6 %)      Full-time n = 81 (21.1 %)<br>Temporary  n = 117 (30.5 %)     Permanent  n = 266 (67.2 %) | |
| Agent n = 43 (7.2 %) | Supervisor n = 342 (87.8 %) |
| **Average Hours and Tenure** | |
| Average Hours Worked per Week = 24.47 | Tenure with company = 3.5 years |

## C. Analyses

To determine which of the independent variables influence the outcome measures of stress, health, performance, and job satisfaction, regression runs were computed for each of these dependent measures. The influence of the two types of control on the outcome measures was also examined.

### 1.  Stress

The ability to adjust the lighting, furniture, and equipment in the work space predicted 7 percent of the reported stress levels. Employees reported significantly more stress at the site with the taller work space panels, and site was a predictor variable of stress. This finding may be due to the small square footage of these work spaces (about 25 square feet) coupled with 72-inch tall panels, creating a potentially claustrophobic "phone booth" environment which also reduces air circulation.  Problems with air movement and odors also predicted stress (and these problems were significantly greater at the site with the tall panels). Together these three variables predicted over 13 percent of the variance in stress complaints.

### 2.  Pain in Hands

Air movement and level of privacy predicted more than 8 percent of the variance in reported pain in arms/hands, and over 12 percent of the variance in reported pain in shoulders and neck.  However, the psychological stress that may result from poor air quality and lack of noise privacy could elevate overall muscle tension. Elevated tension in muscles over prolonged periods can lead to musculoskeletal problems.

## 3. Job Satisfaction

The degree to which the work space layout supports work is the most important predictor of job satisfaction. Work space support of work was measured through satisfaction with amount of work space, arrangement of equipment and furniture, quality of storage, and having to stretch or twist to reach things. It is related to the idea of workspace as a tool for effective work. As an employee supervises more people and gains more computer experience, job satisfaction is reduced. Together, the three variables predicted over 8 percent of the variance in job satisfaction.

## 4. Visual Problems

Air movement and odors predicted almost 8 percent of the variance in visual problems. Poor air quality could lead to eye irritation, or could indirectly cause visual complaints by contributing to a stressful work environment. Lack of control over social interaction, length of employment, and noise problems contribute another 5 percent of the variance in visual problems.

When the variables that are related to musculoskeletal pain were examined, we found in both cases that problems with air quality were the biggest predictors of pain. Other research has suggested the same relationship between air quality and muscle pain. This relationship has been explained like this: Bad air causes psychological stress. Stress causes muscles to tighten up, increases their tension. Prolonged muscle tension, day in and day out, leads to chronic muscle pain. This is how poor air quality can indirectly lead to muscle pain.

Problems with air quality are a predictor of psychological stress. The two main predictors of stress are availability of control over workstation elements and the panel height of the work space. In this case, however, higher panels were associated with greater stress. This may be because the high panels created a "telephone booth" effect, which is unpleasant to be in. This probably contributed to poorer air circulation, and more interruptions by others, because they can't see if anyone is in the workspace.

## D. Design Implications

1. Stress

The most important predictor of stress in this study is the amount of control employees had over lighting, furniture, and computer equipment within their work spaces. It is possible for designers to enhance this aspect of control for clients by specifying furniture that can easily be adjusted or re-configured as the needs of the task at hand change. This study provides evidence that adjustable task lighting and computer support accessories that allow adjustment may reduce the amount of psychological stress experienced by workers. Because stress reactions carry health risks such as musculoskeletal problems and even heart disease, any small modifications to the work environment that reduce stress could have long-term positive impacts on health. Amount of enclosure is a physical element that, in part, determines the amount of control over visual access and social interaction.

## 2. Musculoskeletal Problems

Air quality issues were the most important contributors to musculoskeletal and visual problems. Poor air quality is a stressor. How could bad air lead to muscle pain? There is some evidence to suggest that muscle tension increases with psychological stress, and that over time, high muscle tension can lead to musculoskeletal pain and complaints.

While designers do not specify or design HVAC systems, there are some strategies that can enhance air quality. Most office buildings contain numerous poisons in the air, particularly formaldehyde and carbon monoxide. The incidence of these chemicals can be reduced by growing live plants in the office space. Designing areas within the office space that can hold live plants is an inexpensive way to enhance air quality. Passive airflow can be improved by not bringing panels all the way up to the ceiling, and leaving space above and below panels. Do not place furniture in front of windows. Pay attention to the location of the building air supply and ensure supply and returns are not blocked. Create "aisles" for air flow with furniture layout. To reduce heat retention within the space, specify fabrics (especially in chairs) that do not retain heat. These strategies will enhance air flow and perhaps perceptions of temperature by workers.

## 3. Satisfaction and Performance

This study found that work space layout that properly supports the work process may enhance job satisfaction. With rapid changes in work styles within an organization, effective predesign programming is critical to understanding worker needs and ultimately translating those needs into supportive work space layout. Storage and layout are closely related issues, since both contribute to the efficiency of the work process. Specification of adequate storage capacity, and storage that is appropriate to the job at hand may enhance worker performance.

## IV. CASE STUDY:
## EFFECTS OF AN "ACTIVE STORAGE" WORK
## ENVIRONMENT ON HEALTH AND PRODUCTIVITY OF
## PROJECT MANAGERS AND ACCOUNTANTS

### A. The Project

The purpose of this study was to examine the usefulness of a workstation design concept to support VDT-intensive job types. We examined the effectiveness of this design concept with two different job types: Senior Project Managers (building design and construction responsibilities) and Accountants. These participants worked for two different companies, in different industries, in different areas of North America.

While the functions and responsibilities of these job types varied dramatically from each other, they have in common work processes that included intensive paper-based tasks as well. We evaluated the usefulness of the design concept in terms of how it might enhance health and work effectiveness. Reactions to working in this experimental environment were investigated, including effects on: visual and auditory privacy, psychological stress, musculoskeletal problems, group effectiveness, and individual performance. Other perceptions of support for the VDT work process were also examined, including: the degree to which the workspace aids efficiency, quality of storage, and flexibility and control over pace of work.

Reactions to working in the experimental environment were investigated, including: satisfaction with workspace, visual and auditory privacy, psychological stress, musculoskeletal problems, group effectiveness, and individual performance. Other perceptions of support for the work process were also examined, including: the degree to which the workspace aids efficiency, quality of storage, flexibility of workspace, and control over pace of work.

### B. Problem Statements

The research team developed a number of problem statements around which to articulate the design solution. These problem statements are generalizable to a wide variety of work types but are closely related to the problems of concurrent VDT and paper media usage.

Paper is now a display medium in addition to being a storage medium. It is used to cue people's work, remind them of tasks, and order their work. Because of this use, people want paper-based materials kept within visual range. Thus people tend to keep piles of paper representing work-in-process on the desk top.

A significant amount of paper-based material (articles, magazines, memos) "passes through" a workspace but has no place to rest while it is there.

Computers and related peripherals have taken up some 10 to 15 percent of the workspace that was formerly used for work-in-process support. There is less work surface and storage available to handle ongoing work. Thus, materials related to work-in-process are concentrated into a small space.

The pace of work is increasing, reducing time available to file, properly store, or otherwise organize work materials. Workers may feel compelled to pile materials on the work surface and floor, rather than in files.

With the increase in pace, workers are under more pressure to have easy access to important information. This also encourages the tendency to pile materials within the workspace.

## C. Design Concept Response

The design of the experimental workspace concept addressed two general criteria: 1) support of simultaneous VDT and paper based tasks within the workstation; 2) support for visual and acoustic privacy through physical enclosure. Figure 7.6 is an illustration of the design concept response to the problem statements and criteria.

The central feature of the design is the integration of a storage unit that also permits the display of paper materials that are in active use. The storage unit also takes the place of a workstation panel.

## D. Research Hypotheses

The primary focus of this study was to test how well the design concept supported the health and effectiveness of workers in constrained spaces performing VDT and paper-based tasks. A set of research questions was developed to respond to these problem statements.

### 1. Hypothesis 1: Individual and Group Performance

The design concept is intended to facilitate enhanced organization and storage of work materials that may result in performance gains. Because the active storage concept can be used to create workspaces for work teams and work groups, it is hypothesized that it will also enhance the effectiveness of teams because it increases the level of organization of work materials, provides better storage, and enhances privacy through better enclosure.

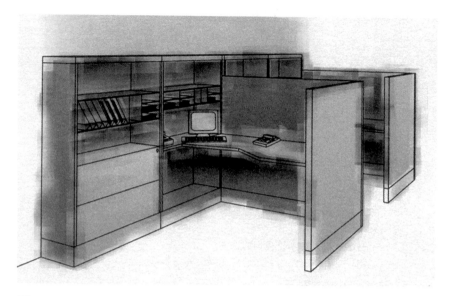

**Figure 7.6** Design concept (With permission. Copyright 1993, Herman Miller, Inc., All Rights Reserved.)

## 2. Hypothesis 2: Privacy

The design concept was intended to provide a means of visually dividing space and providing enclosure. Because the cabinet depth is much greater than the systems furniture panels it is designed to replace, perceptions of acoustic and visual privacy should be enhanced for workers within this environment. Enhanced privacy may be related to reduced stress levels for workers.

## 3. Hypothesis 3: Flexibility and Control

The design concept was intended to enhance control over the workspace (and the work process) by facilitating the organization and display of information critical to effective completion of job-related tasks. Thus, the experimental environment may enhance employee control over the work process by improving control over the workspace. It is also hypothesized that enhanced control will result in a decrease in psychological stress and increase the productivity of workers within the experimental workspace.

## 4. Hypothesis 4: Muscle Pain

The 90-degree relationship between the storage cabinet and the work surface will reduce the amount of reaching and twisting by the user for needed materials. Thus we hypothesized that the use of this experimental workspace would reduce complaints of musculoskeletal pain.

## 5. Hypothesis 5: Stress

It is hypothesized that enhancements to visual and acoustic privacy from use of this workspace design concept would result in reductions in psychological stress.

## E. Settings and Participants

We examined the effectiveness of this design concept with two different job types: Senior Project Managers (building design and construction responsibilities) and Senior Accountants. These participants worked for two different companies, in different industries, in different geographic areas of North America.

While the functions and responsibilities of these job types varied dramatically from each other, they have in common work processes that included intensive computer use and paper-based tasks as well.

When the workstations were redesigned to accept the experimental workspaces for this study, they were kept to the same or slightly smaller square footage than the original workspaces. This was done to prevent jealousy on the part of co-workers at each site not receiving new workstations, and to prevent problems with user evaluations that may have occurred with an increase in workspace square footage. In both test locations, workers were already housed within systems furniture cubicles.

## F. Research Design

This is a longitudinal study in which user perceptions were measured at three points in time over the duration of the project. At each of the two study sites, the perceptions of a Control group of employees in the same company performing identical work in workstations without the experimental design concept were also assessed. The first data collection occurred 2 months before the work environments were modified. The second data collection occurred about 1 month after the design intervention. The final data collection point was 4 months after the change to the work environment.

The data was collected over several points in time because research has shown that when people are being studied they tend to report positive increases in performance and satisfaction regardless of whether or not the intervention was actually effective. This is known as the "Hawthorne Effect." Collecting data at several points in time over a long period reduces this problem because participants grow accustomed to being studied. In addition, use of Control group data permits any Hawthorne effects, or the effects of outside influences on test results (such as organizational changes in the company or the state of the national economy) to be statistically identified and controlled.

## G. Senior Project Managers

### 1. Participants

The participants were four assistant vice president or director-level executives (three males, one female) whose primary responsibilities are to manage the planning, design, remodeling, and construction of retail and corporate facilities. These workers manage projects and meet with external clients, vendors, and cross functional internal team members. The members of this group seldom interacted with each other on projects, but they are formally identified as a group.

### 2. Physical Environment

These workers have unusual paper storage needs because much of the paper they handle is oversize (such as blueprints, legal contracts, and construction documents). Figure 7.7 shows a photograph of a typical workstation for this job type. Note that there is a work surface, a table for meetings (and paper display), storage for a variety of documents in overhead compartments, and tackable space. There is also space for a computer and monitor (not shown).

At the same time the Experimental group was identified, a Control group of participants was created, having identical job descriptions and workstations. The participants in this group were evaluated on all the same measures of interest to this study, but received no change to their work environments.

Figure 7.8 shows a photograph of a typical experimental workstation developed for workers in the test group. In addition to receiving the storage cabinets integrated into their work surfaces, this group received freestanding height-adjustable desks. The work surface in each workstation had a "cutaway" portion. The rationale for the cutaway is to create a "cockpit" that the user can pull up into, thus providing easier reach into the cabinet and to the computer or other items placed on the desk.

Figure 7.9 shows a rough drawing of the layout of the four workspaces that we modified for this study. We were not trying to create a team space for this project because these individuals did not work in that fashion. Rather, we were trying to redesign the work spaces to roughly fit in the same footprint, or smaller than the workspaces that they were replacing.

**Figure 7.7** Typical workstation prior to change (Author)

## 3. Methods

A questionnaire examined perceptions of control over pace of work, stress, musculoskeletal complaints, individual performance, group effectiveness, and privacy.

Evaluations of the workspace were also recorded, including: overall satisfaction, quality of storage, degree to which workspace aids efficiency and organization of materials, and flexibility/control over workspace. Some of these questions have been developed and used in previous research (Brill, 1984; Caplan et al., 1975; Nadler, 1977; O'Neill, 1992).

Multiple-item measures of these constructs were created and the reliability of these measures tested using Chronbach's alpha ($\alpha$). This statistic allows researchers to ask a number of questions relating to the same concept (for example, privacy) and then statistically collapse the responses to these questions into a single index representing that construct.

**Figure 7.8** Typical workstation layout for Experimental group (Author)

Chronbach's alpha can range from 0 (representing a completely unreliable index) to 1 (meaning a highly reliable and useful index). Thus, the greater the alpha, the more reliable the scale.

*Individual Performance.* Individual performance was measured through eight items (each using a 9-item Likert-type scale). The scale asked people to evaluate how well they do their job within a range from 1 = unacceptable, 5 = reasonably well, to 9 = perfect/ideal. The individual ratings included: amount of work accomplished, quality of work, error rate, taking responsibility, creativity, getting along with others, dependability, and overall performance (Brill, 1984). These items were averaged to form an index ($\alpha = .86$).

*Group Effectiveness.* Each participant was asked to rate the effectiveness of their work group on seven dimensions (each using a 7-item Likert-type scale). On this rating scale, 1 = ineffective and 7 = effective. The following items were averaged to form an index ($\alpha = .91$): problem solving; making decisions; getting work done; using member skills, abilities, and resources; meeting individual needs; satisfaction with group membership (the scale anchors for this item are 1 = very dissatisfied to 7 = very satisfied); and overall effectiveness of group (Nadler, 1977).

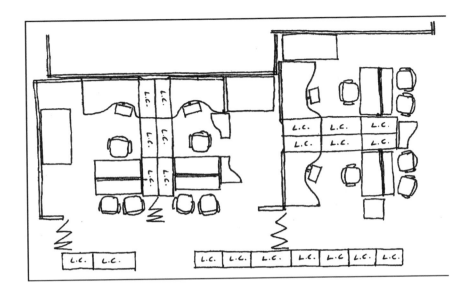

**Figure 7.9** Layout of four workspaces in Experimental group (With permission. Copyright 1993, Herman Miller, Inc., All Rights Reserved.)

***Stress.*** To measure psychosomatic symptoms of stress, eight items measuring this construct (on a 3-point Likert scale) were averaged to form an index ($\alpha$ = .69): heart beating hard, shortness of breath, dizziness, hands trembling, upset stomach, heart beating fast, ill health, and loss of appetite (Caplan et al., 1975).

***Workspace Aids Efficiency.*** Each participant was asked to rate the degree to which their workspace enhances their efficiency. To measure workspace efficiency, four items measuring this construct were averaged to form an index ($\alpha$ = .61): workspace helps me get my job done efficiently; finding work materials quickly; repeating tasks unnecessarily (reverse scored); and losing track of where I am in a task (reverse scored).

***Quality of Storage.*** To measure storage quality, three items measuring this construct (on a 5-point Likert scale) were averaged to form an index (Pearson's r = .57 to .74 between variables): satisfaction with amount of storage, number of file drawers, and number of shelves.

***Stretching and Twisting.*** To measure problems with users having to stretch and twist while reaching for work materials within the workspace, six items measuring this construct (on a 5-point Likert scale) were averaged to form an index ($\alpha$ = .87): bothered by having to stretch or twist, working in uncomfortable positions, awkward work motions, stretching or twisting to reach work materials, having to reach too far while seated, and having to twist too far while seated.

*Privacy.* Separate scales were developed for visual and auditory privacy. Three items measuring auditory privacy were developed (using a 5-point Likert-type scale) and averaged to form an index (Pearson's $r$ = .48 to .86 between variables): ability to have confidential conversations in workspace (reversed), overhearing immediate neighbors, and being overheard from the workspace. Four items measuring visual privacy (on a 5-point Likert-type scale) were averaged to form an index ($\alpha$ = .80): distracted by seeing others, interrupted while working, too exposed to view of others, and satisfaction with privacy (reversed).

*Control over Organization of Workspace.* Each participant was asked to rate the degree to which the workspace supports the rearranging and organization of work. Five items measuring this construct (on a five-point Likert scale) were averaged to form an index ($\alpha$ = .92): ability to arrange work materials to match work style, satisfaction with arrangement of equipment, ease of organizing work materials, ability to rearrange work materials to suit the task at hand, and amount of flexibility/adjustability in workspace.

*Control over Pace of Work.* Control over pace of work was measured by a single item (measured on a 5-point Likert-type scale): "feeling in control of the pace of my work."

*Satisfaction with Workspace.* This variable was measured through a single item (measured on a 5-point Likert-type scale): "In general, I am satisfied with my workspace."

*Musculoskeletal Pain.* Five aspects of musculoskeletal pain were measured on a 3-point Likert-type scale (1 = never, 3 = often), including: back pain, pain in arms and hands and fingers, pain in neck and shoulders, pain in legs, and general muscle pain. These indexes were created from questionnaire items developed by Sainfort (1990).

*Visual Problems.* Visual problems were assessed through items measured on a 3-point Likert-type scale (1 = never, 3 = often). These items assessed frequency of headaches, burning eyes, tearing or itching eyes, and eye strain or sore eyes. These indexes were created from questionnaire items developed by Sainfort (1990).

## 4. Analyses

This section describes the analyses conducted for this study. This research was originally designed as a longitudinal study, with data collected from a Control group and from the Experimental group at three points in time over a six-month period. Although data was collected in this manner at each site, there were several problems with the sample. Project time constraints (the entire project was initially delayed several months) prevented Control group data from being collected at the third data collection point. There is a small total sample size and a relatively small number of participants, resulting in small cell sizes for a longitudinal analysis.

Because of these constraints, it was decided to group all Control responses into one group and all responses from participants in the Experimental condition into another group for purposes of analysis.

Analyses of Variance (ANOVAs) confirmed that there were no significant differences in responses to the major variables of interest between the second and third data collection points for the Experimental group. Thus it was possible to combine the responses from these two groups. There were no significant differences across the three data collection points on the Control groups in the three settings, so it was possible to collapse all Control group data into one group. The fact that the responses of the Experimental and Control groups were stable across time suggests that the questionnaire is a reliable instrument. These steps have the positive result of greatly simplifying the subsequent analyses and interpretation of results. However, because of the number and complexity of analyses, only statistically significant findings are reported in this section.

*t-Tests for Differences between Experimental and Control groups.* The first set of analyses for each site consists of t-tests computed to discover differences between the reported means on the outcome measures (for example, on privacy and performance) between the Control and Experimental groups. The results of this analysis are shown in Table 7.5. This analysis determines statistically significant differences between the groups on the outcome measures but does not indicate the cause of the differences.

Participants using the active storage design concept reported significantly greater individual performance ratings than did the Control group (Control M = 7.42, Exp. M = 7.89, t = 2.34, df = 1, 16, p < .05, 9-point scale) and felt their group effectiveness was significantly greater than did those in the Control group (Control M = 5.21, Exp. M = 6.05, t = 2.53, df = 1, 13, p <.05). Users of the active storage concept reported significantly greater visual privacy than the Control group (Control M = 3.3, Exp. M = 2.5, t = 1.87, df = 1, 15, p = .08). There were no other significant differences in perceptions between the two groups.

Results of these comparisons show that employee evaluations of performance, group effectiveness, and visual privacy are significantly more positive for participants in the Experimental group. These results do not explain *why* workers in the Experimental group scored higher on the outcome measures, such as performance or stress.

***Multiple Regressions.*** Multiple regressions were computed to determine

| Table 7.5 Significant Differences Between Means of Control and Experimental groups (Project Managers) | | | | |
|---|---|---|---|---|
| | **Control Group Mean** | **Experi-mental Group Mean** | **t value** | **Signifi-cance Level** |
| Performance | 7.42 | 7.89 | 2.34 | p < .05 |
| Group Effec-tiveness | 5.21 | 6.05 | 2.35 | p < .05 |
| Visual Privacy[1] | 3.30 | 2.50 | 1.87 | p = .08 |

[1] A numerically greater mean indicates greater problems with privacy

the design features or other variables that predict user responses. For example, if we find significant differences in privacy between the Experimental and Control groups, what design features of the workspace are predictors of privacy? Through multiple regression, we may be able to determine that, for instance, panel height is the greatest predictor of privacy. Thus, in this case, we would know that the difference in privacy between the two groups was related to panel height, not to the use of the active storage design concept.

The predictive effects of many of the study variables were examined. To examine the overall effects of the active storage concept, participants from the Control group were assigned a score of 0 and participants from the Experimental group were assigned a score of 1. This new variable, "Active Storage use," was used as a "dummy" variable in the regression analyses.

Multiple regression information is presented in tables throughout the text. Each table uses several headings. The "$R^2$" heading indicates the total amount of variance in the outcome measure predicted by all predictor variables. The size of the numbers under the "Standardized Coefficient" heading indicates the relative size of the contribution of each predictor variable (if there is more than one) to the variance in the outcome measure. The "F" heading is the F value for that variable. A higher F value indicates a greater level of statistical significance for the predictor variable. The "p" heading is the probability of error (chances in 100) in stating that the relationship between the two variables is statistically significant. Thus, if p = .02, there is a 2 percent chance that the relationship is not really statistically significant.

In order to determine the variables that predict privacy and performance for Project Managers, regression analyses were computed. These analyses show the relative contributions of each predictor variable on employee reactions. Table 7.6 shows a list of the outcome measures, with the significant predictor variables listed below in descending order of relative importance.

Table 7.6 shows that the variable, "Active Storage," is the most significant predictor of individual performance, group performance, and visual privacy for these workers. Because the t-tests show that workers having the integrated storage concept scored significantly higher on these three outcomes, the results suggest that use of this concept is a predictor of these enhanced outcomes. Other design-related variables such as control over organization of materials, the degree to which the workspace aids efficiency, and auditory privacy were also predictors of the outcome measures.

| Table 7.6 Regression Analyses (Project Managers) | | | | | |
|---|---|---|---|---|---|
| **Outcome** | **$R^2$** | **Predictor Variables** | **Beta coefficient** | **F** | **p** |
| Performance | .38 | - Active Storage | .51 | 2.50 | .02 |
| | | - Efficiency of Workspace | .39 | 1.90 | .07 |
| Group Effectiveness | .69 | - Active Storage | .70 | 3.94 | .00 |
| | | - Control over Organization | .72 | 3.89 | .03 |
| | | Auditory Privacy | .46 | 2.48 | .03 |
| Visual Privacy | .19 | - Active Storage | .43 | 1.87 | .08 |

Table 7.6 shows that use of Active Storage, and efficiency of workspace, predict 38 percent of the variance in Project Manager performance. Active storage, control over organization and flexibility of materials, and auditory privacy, predict 69 percent of the variance in the effectiveness of work groups. Use of the Active storage concept was the sole predictor of visual privacy, accounting for 19 percent of the variance in perceptions of privacy for Project Managers. Comments derived from user interviews over the course of the study support the empirical findings:

"I'm more organized than before. It's easier to find things."

"I find it easier to work."

"I'm faster and more efficient... I've moved paper around the pockets (in the Active storage concept) over time."

## H. ACCOUNTANTS

### 1. Participants

The company is a large, multinational accounting firm. The participants are high-level accountants (two men, two women) whose jobs are structured with a great degree of autonomy. These workers manage individual accounts with long-term clients. They have an extremely heavy workload from January through April. They do not interact with each other on project work. They are, however, part of a larger, identified group of accountants within the organization.

### 2. Physical Environment

These accountants work with a high volume of standard and legal size paper that is held in large cardboard boxes in archival storage. These boxes are brought directly into the workspace as needed. The boxes generally stay in the workspace until the project is complete (which may be 2 to 3months). Figure 7.10 shows a photograph of a typical Accountant's workstation.

Each workstation has two work surfaces, which are typically covered with papers. In addition, each workstation contains a storage unit with two lateral files. This is also often used as a place to stack papers for visual reference (see Figure 7.10). The workstations do not have space for meetings with others. Meetings are held in other spaces in the area. Most of the work of these accountants is solitary in nature, with the exception of times when they may be conferring with colleagues or clients.

At the same time the Experimental group was identified, a Control group of participants was created, having identical job descriptions and workstations. The participants in this group were evaluated on all the same measures of interest to this study, but received no change to their work environments.

Figure 7.11 shows a photograph of a typical experimental workstation developed for workers in the test group. In addition to receiving the storage cabinets integrated into their work surfaces, this group received freestanding height-adjustable desks. The work surface in each workstation had a "cutaway" portion. The rationale for the cutaway is to create a "cockpit" that the user can pull up into, thus providing easier reach into the active storage cabinet and to the computer or other items placed on the desk.

Participating employees in the Control group used standard systems furniture products.

Figure 7.12 shows a rough drawing of the layout of the four workspaces that we modified for this study. We were not trying to create a team space for this project because these individuals did not work in that fashion. Rather, we were trying to redesign the work spaces to roughly fit in the same footprint, or smaller than the workspaces that they were replacing.

## 3. Methods

A questionnaire examined perceptions of: control over pace of work, stress, musculoskeletal complaints, individual performance, group effectiveness, and privacy.

**Figure 7.10** Typical accountant workstation before change (Author)

Evaluations of the workspace were also recorded, including: overall satisfaction, quality of storage, degree to which workspace aids efficiency and organization of materials, and flexibility/control over workspace. Some of these questions have been developed and used in previous research (Brill, 1984; Caplan et al., 1975; Nadler, 1977; O'Neill, 1992).

Multiple-item measures of these constructs were created and the reliability of these measures tested using Chronbach's alpha ($\alpha$). This statistic allows researchers to ask a number of questions relating to the same concept (for example, privacy) and then statistically collapse the responses to these questions into a single index representing that construct. Chronbach's alpha can range from 0 (representing a completely unreliable index) to 1 (meaning a highly reliable and useful index). Thus, the greater the alpha, the more reliable the scale.

**Figure 7.11** Experimental workstation with active storage and cut-away work surface (Author)

*Individual Performance.* Individual performance was measured through eight items (each using a 9-point Likert-type scale). The scale asked people to evaluate how well they do their job within a range from 1 = unacceptable, 5 = reasonably well, to 9 = perfect/ideal. The individual ratings included: amount of work accomplished, quality of work, error rate, taking responsibility, creativity, getting along with others, dependability, and overall performance (Brill, 1984). These items were averaged to form an index ($\alpha = .86$).

***Group Effectiveness.*** Each participant was asked to rate the effectiveness of their work group on seven dimensions (each using a 7-item Likert-type scale). On this rating scale, 1 = ineffective and 7=effective. The following items were averaged to form an index ($\alpha = .91$): problem solving; making decisions; getting work done; using member skills, abilities, and resources; meeting individual needs; satisfaction with group membership (the scale anchors for this item are 1=very dissatisfied to 7 = very satisfied); and overall effectiveness of group (Nadler, 1977).

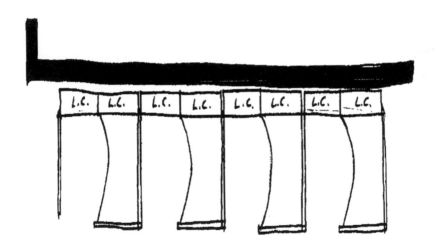

**Figure 7.12** Layout of four workspaces in Experimental group (With permission. Copyright 1993, Herman Miller, Inc., All Rights Reserved.)

***Stress.*** To measure psychosomatic symptoms of stress, eight items measuring this construct (on a 3-point Likert scale) were averaged to form an index ($\alpha = .69$): heart beating hard, shortness of breath, dizziness, hands trembling, upset stomach, heart beating fast, ill health, and loss of appetite (Caplan et al., 1975).

***Workspace Aids Efficiency.*** Each participant was asked to rate the degree to which their workspace enhances their efficiency. To measure workspace efficiency, four items measuring this construct were averaged to form an index ($\alpha = .61$): workspace helps me get my job done efficiently; finding work materials quickly; repeating tasks unnecessarily (reverse scored); and losing track of where I am in a task (reverse scored).

*Quality of Storage.* To measure storage quality, three items measuring this construct (on a 5-point Likert scale) were averaged to form an index (Pearson's r = .57 to .74 between variables): satisfaction with amount of storage, number of file drawers, and number of shelves.

*Stretching and Twisting.* To measure problems with users having to stretch and twist while reaching for work materials within the workspace, six items measuring this construct (on a 5-point Likert scale) were averaged to form an index (α = .87): bothered by having to stretch or twist, working in uncomfortable positions, awkward work motions, stretching or twisting to reach work materials, having to reach too far while seated, and having to twist too far while seated.

*Privacy.* Separate scales were developed for visual and auditory privacy. Three items measuring auditory privacy were developed (using a 5-point Likert-type scale) and averaged to form an index (Pearson's r = .48 to .86 between variables): ability to have confidential conversations in workspace (reversed), overhearing immediate neighbors, and being overheard from the workspace. Four items measuring visual privacy (on a 5-point Likert-type scale) were averaged to form an index (α = .80): distracted by seeing others, interrupted while working, too exposed to view of others, and satisfaction with privacy (reversed).

*Control Over Organization of Workspace.* Each participant was asked to rate the degree to which the workspace supports the rearranging and organization of work. Six items measuring this construct (on a five-point Likert scale) were averaged to form an index (α = .92): ability to arrange work materials to match work style, satisfaction with arrangement of equipment, ease of organizing work materials, ability to rearrange work materials to suit the task at hand, and amount of flexibility/adjustability in workspace.

*Control Over Pace of Work.* Control over pace of work was measured by a single item (measured on a 5-point Likert-type scale): "feeling in control of the pace of my work."

*Satisfaction with Workspace.* This variable was measured through a single item (measured on a 5-point Likert-type scale): "In general, I am satisfied with my workspace."

*Musculoskeletal Pain.* Five aspects of musculoskeletal pain were measured on a 3-point Likert-type scale (1 = never, 3 = often), including: back pain, pain in arms, hands, and fingers, pain in neck and shoulders, pain in legs, and general muscle pain. These indexes were created from questionnaire items developed by Sainfort (1990).

*Visual Problems.* Visual problems were assessed through items measured on a 3-point Likert-type scale (1 = never, 3 = often). These items assessed frequency of headaches, burning eyes, tearing or itching eyes, and eye strain or sore eyes. These indexes were created from questionnaire items developed by Sainfort (1990).

## 4. Analyses

This research was originally designed as a longitudinal study, with data collected from a Control group and from the Experimental group at three points in time over a six-month period. Although data was collected in this manner at each site, there were several problems with the sample. Project time constraints (the entire project was initially delayed several months) prevented Control group data from being collected at the third data collection point. There is a small total sample size and a relatively small number of participants in each site, leading to small cell sizes for a longitudinal analysis.

Because of these constraints, it was decided to group all Control responses into one group and all responses from participants in the Experimental condition into another group for purposes of analysis.

Analyses of Variance (ANOVAs) confirmed that there were no significant differences in responses to the major variables of interest between the second and third data collection points for the Experimental group. Thus it was possible to combine the responses from these two groups. There were no significant differences across the three data collection points on the Control groups in the three settings, so it was possible to collapse all Control group data into one group. The fact that the responses of the Experimental and Control groups were stable across time suggests that the questionnaire is a reliable instrument. These steps have the positive result of greatly simplifying the subsequent analyses and interpretation of results. However, because of the number and complexity of analyses, only statistically significant findings are reported in this section.

***t-tests for Differences between Experimental and Control groups.*** The first set of analyses for each site consists of t-tests computed to discover differences between the reported means on the outcome measures (for example on privacy and performance) between the Control and Experimental groups. This analysis determines statistically significant differences between the groups on the outcome measures but does not indicate the cause of the differences.

Participants using the active storage cabinet as part of a workstation design reported significantly greater individual performance ratings than did the Control group (Control M = 6.98, Exp. M = 7.70, t = 2.34, df = 1, 21, p = .06. Note that the mean scores are on a 9-point scale). Participants using the cabinet reported a significantly greater ability to rearrange and organize work materials than did workers in the Control group (Control M = 2.77,

Exp. M = 3.83, t = 2.524, df = 1, 21, p < .01). There were no other significant differences in perceptions between the two groups.

Table 7.7 shows that employee evaluations of performance and control over organization of work materials are significantly more positive for workers having the Active storage cabinet. These results do not explain why these workers scored higher on the dependent measures.

| Table 7.7 Significant Differences Between Experimental and Control groups (Accountants) | | | | |
|---|---|---|---|---|
| | **Control group Mean** | **Experi-mental Group-Mean** | **t value** | **Significance Level** |
| **Performance** | 6.98 | 7.70 | 2.34 | p = .05 |
| **Control over Organization** | 2.77 | 3.83 | 2.52 | p < .01 |

***Multiple Regressions.*** The t-test analyses are followed by multiple regressions computed to determine the design features or other variables that may predict user responses. For example, if we find significant differences in privacy between the two groups, what design features of the workspace are predictors of privacy? Through multiple regression, we may be able to determine that, for instance, panel height is the greatest predictor of privacy. Thus, in this case we would know that the difference in privacy between the two groups was related to panel height, not to use of the Active storage design concept.

The predictive effects of many of the study variables were examined. To examine the overall effects of Active Storage use, employees from the Control group were assigned a score of 0 and participants from the Experimental group were assigned a score of 1. This new variable, "Active Storage use," was used as a "dummy" variable in the regression analyses.

Multiple regression information is presented in Table 7.7. The "$R^2$" column heading indicates the total amount of variance in the outcome measure predicted by all predictor variables. The size of the numbers under the "Standardized Coefficient" heading indicates the relative size of the contribution of each predictor variable to the variance in the outcome measure. The "t" heading is the tvalue for that variable. A higher tvalue indicates a greater level of statistical significance for the predictor variable. The "p" heading is the probability of error (chances in 100) in stating that the relationship between the two variables is statistically significant. Thus, if p = .02, there is a 2 percent chance that the relationship is not really statistically significant.

In order to determine the variables that predict control over organization of work materials and performance at this site, regression analyses were computed. These analyses show the relative contributions of each predictor variable on employee reactions. Table 7.8 shows a list of the outcome measures (leftmost column), with the significant predictor variables listed (column titled "Predictor Variables") in descending order of relative importance.

| Table 7.8 Regression Analyses (Accountants) | | | | | |
|---|---|---|---|---|---|
| **Outcome** | **R²** | **Predictor Variables** | **Beta Coefficient** | **T** | **p** |
| **Performance** | .32 | Efficiency of workspace | .42 | 2.16 | .04 |
| | | Active Storage | .30 | 1.52 | .10 |
| **Control over Organization** | .62 | Filing | .45 | 5.62 | .000 |
| | | Efficiency of workspace | .33 | 3.99 | .000 |
| | | Auditory Privacy | .17 | 2.13 | .03 |
| | | Control over Work Pace | .16 | 2.12 | .03 |
| | | Active Storage | .12 | 1.56 | .12 |
| **Stress** | .20 | Control Work Pace | .45 | 2.26 | .05 |

Table 7.8 shows that the degree to which the workspace enhances efficiency is the most significant predictor of individual performance, followed by the use of the Active storage concept. Together, these variables predict 32 percent of the variance in performance. Although overall workspace efficiency predicts performance, the use of Active storage does not directly predict increased workspace efficiency.

Rather, the results suggest that *any* workspace that people report as making them more efficient would lead to greater individual performance. We did find a direct relationship between use of Active storage and performance.

Table 7.8 shows that assessed quality of filing was the most powerful predictor of the amount of control over organization of work materials a workspace offers (control over rearranging furnishings, equipment, and paperwork to fit the task at hand). Perceived efficiency of the workspace is the next greatest predictor of control over workspace. Auditory privacy, control over the pace of work, and type of furniture used were additional predictors of control. The five variables together predict 62 percent of the variance in control over organization of work materials (see Table 7.8).

Because workers in the experimental workstations did not rate their workspaces significantly better on storage, efficiency, auditory privacy, or control over pace of work, these variables are not linked to any of the design concepts we tested. Rather, this finding shows that, in general, organization of work materials is related to storage, efficiency of workspace, and control over pace of work. Use of the Active storage concept predicted a small amount of the variance in "control over organization of work materials" scores (see Table 7.8).

Table 7.8 shows that control over the pace of work is the only predictor of psychological stress symptoms, accounting for 20 percent of the variance in reported stress.

We were interested in gaining a better understanding of the relationship between the Active storage concept and stress in these professional workers. We conducted several additional regression analyses to create a path analytic model of this relationship. This complex model shows a relationship between use of the Active storage concept, along with other intervening variables, and stress reductions in these workers.

We found that the amount of control over the organization of work materials in the workspace predicts 10 percent of the variance in perceptions of control over the pace of work. Availability of Active storage, adequate filing, and efficiency of workspace predicts enhanced control over organization of work (see Table 7.8). In turn, we found that control over organization of work predicts perceived control over the pace of work. Finally, our analyses revealed that enhanced control over the pace of work was related to significant stress reductions for these professional accountants. Figure 7.13 illustrates this complex relationship.

**Active Storage Concept = = > Enhanced Organization of Work Materials = = > Enhanced Control over Work Pace = = > Reduced Stress**

**Figure 7.13** Path model of active storage and stress (Author)

## 5. Conclusions

*Hypothesis 1: Individual and Group Performance.* The experimental design concept is designed to facilitate the organization of work-in-process, including storage of paper and work tools. Enhanced organization and storage in work materials may result in performance gains. Because the storage design concept can be used to create workspaces for work teams and work groups, it was hypothesized that use of this concept would also enhance the effectiveness of teams because it increases the level of organization of work materials, provides better storage, and enhances privacy through better enclosure than panels alone can provide.

Self-rated, individual performance significantly improved for workers in both of the test sites. Analyses showed that these performance gains were related to use of the Active storage concept, and a second design variable, the degree to which the workspace enhances efficiency.

Group effectiveness, a measure of team or group performance, significantly increased for Project managers using the experimental workstations. Analyses showed that this gain in group performance was predicted by three variables: use of the Storage concept, degree of organization over work materials provided by the workspace, and perceived auditory privacy.

*Hypothesis 2: Privacy.* The Active storage design concept was intended to provide a means of visually dividing space and providing enclosure. Because the cabinet depth is much greater than the systems furniture panels it is designed to replace, perceptions of acoustic and visual privacy should be enhanced for users of the experimental workstation, due to a greater feeling of enclosure. Enhanced privacy may be related to reduced stress levels for workers.

Visual privacy significantly increased for Project managers. Analyses showed that the gain in visual privacy in both sites was predicted by one variable: use of the Active storage concept.

*Hypothesis 3: Flexibility and Control.* The work environment can be designed to enhance control over the job. The Storage concept may enhance employee control over the work process by improving control over the workspace. The storage concept is intended to enhance control over the workspace by facilitating the organization and display of information critical to effective completion of job-related tasks. It was hypothesized that enhanced control (provided through use of the Active storage concept) would lead to a decrease in psychological stress and increase the effectiveness of workers using the workspace.

Control over organization of work materials (ability to arrange materials to match work style, ease of organizing materials, rearranging to match the task at hand, and flexibility/adjustability of workspace) was significantly greater for Accountants using the experimental workstations. Analyses showed that control was predicted by three main variables: quality of filing (satisfaction with amount of storage, number of file drawers, and number of shelves), degree to which the workspace supports efficient design, and use of the Active storage concept. No decrease in psychological stress from greater control was found.

*Hypothesis 4: Muscle Pain.* The 90-degree relationship between the cabinet and the work surface will reduce the amount of reaching and twisting by the user for needed materials. Thus the use of this product should reduce complaints of musculoskeletal pain.

No differences in problems with stretching or twisting were found between the Experimental and Control groups. No differences in musculoskeletal pain were found between the Experimental and Control groups.

*Hypothesis 5: Stress.* It was hypothesized that enhancements to privacy from use of the experimental workstations would result in self-reported reductions in psychological stress.

Analyses revealed an indirect relationship between use of the storage design concept, some design characteristics of the workstation, and stress reductions in professional Accountants. Use of the Active storage concept and other characteristics (storage and efficiency of workspace) enhance control over organization of work. In turn, control over organization of work predicts perceived control over the pace of work for this group. Finally, enhanced control over the pace of work was related to significant stress reductions for these professionals. This complex relationship looks like the illustration in Figure 7.13.

Some initial reactions to stressful situations at work are heart pounding, shortness of breath, upset stomach, and loss of appetite. While these symptoms themselves are not serious, over time, stress can lead to serious health problems such as hypertension and coronary heart disease. Any design feature that can reduce stress complaints has the potential to positively affect workers' long-term health.

## V. CASE STUDY: EFFECTS OF WORKSTATION ADJUSTABILITY AND TRAINING ON STRESS AND MOTIVATIONAL PERFORMANCE

### A. Background

Many studies have demonstrated that giving people more control over decisions affecting their work can enhance physical health and offset the negative effects of such conditions as high workload or fast-paced work.

Considering that the benefits of providing control over decision-making, pacing, and design of work tasks are well established, surprisingly little research has focused on the physical environment as an element of control. Therefore this study was designed to learn more about the effects of giving people enhanced control over their immediate physical environment at work.

The purpose of this laboratory experiment was to examine the effects of interior workstation adjustability - and the effects of training in how to use that adjustability - on physiological stress and motivational performance levels under high workload. We hypothesized that more control (through adjustability) and training would have the most desirable impact on these outcomes. (This study originally appeared in the EDRA 2000 Proceedings, O'Neill, M. and Evans, G., authors, and is reprinted with permission of EDRA, copyright 2000, All Rights Reserved).

### B. Introduction

Control -- the ability to alter the pace of work, to order tasks, and to make decisions on the job -- has a significant impact on health and well being in the workplace (Karasek and Theorell, 1990). Many studies have demonstrated that giving people more control over decisions affecting their work can enhance physical health and offset the stressing effects of such conditions as high workload or fast pace of work (Evans and Cohen 1987; Jackson, 1989; Sauter, Hurrell, and Cooper 1989).

Considering that the benefits of providing control over decision-making, pacing, and design of work tasks are well established, surprisingly little research has focused on the physical environment as an element of control. In addition, while the concept of stress is discussed in the literature on office design research, there has been little investigation of the relationship between design features and stress. However, O'Neill (1998) has reported on a series of studies indicating a relationship between workstation adjustability and self reported stress and work performance among office workers.

Over the last decade, studies have shown a causal link between elevated levels of cortisol and epinephrine in the blood and coronary heart disease (Baum and Grunberg 1995; Grunberg and Singer 1990). However, in work environment studies, stress is typically measured through a "paper and pencil" test that probes respondents' feelings. These psychological measures are suggestive, but physiological reactions to stressful events, particularly the release of certain types of hormones into the blood, have more reliable implications for stress reduction and associated health outcomes for office workers.

To learn more about the effects of giving people enhanced control over their immediate physical environment at work, the authors conducted a controlled laboratory experiment. The experiment attempted to quantify the relationship between specific work environment design characteristics and physiological stress and motivation reactions in office employees under high workload conditions.

## C. The Experiment

The purpose of the experiment was to examine the effects of interior workstation adjustability -- and the effects of ergonomic training in how to use that adjustability -- on stress and motivation levels under high workload. The central focus of this study was to test the notion that increased control through physical adjustability and training would have the greatest impact on these outcomes.

### 1. The Work Environment

The experimenters set up two workstations to simulate an office work environment. One was a "typical" cubicle formed by fixed partitions supporting a 30" X 60" work surface (see Figure 7.14). This workstation was designed to represent the degree of adjustability typically found in corporate offices used by a wide variety of knowledge work job types across numerous industries. In this typical cubicle, the computer monitor and keyboard could be repositioned, the chair could be moved, and the hanging display storage and white board/tack board could be repositioned. The other workstation was a prototype designed to maximize options for internal adjustment. The standard cubicle is shown in Figure 7.14; two views of the prototype work-station are shown in Figure 7.15.

**Figure 7.14** Standard cubicle used in the study (Author)

2. The Work Task

The primary activity during the 3 hour work period was typing a manuscript off of hard copy. The manuscript was a technical report on an unfamiliar topic. Periodically, common office clerical tasks were interspersed and briefly interrupted the manuscript task.

Participants expected these interruptions and were told that performance was not of interest during the experiment and that they should try to work at a normal work pace on the tasks. Every 30 minutes, an additional clerical task was introduced by the experimenter. The tasks were designed both to represent typical office work and to encourage both changes in posture and movement at the work station.

3. Measures

During the 3 hour work period, a trained observer unobtrusively monitored each participant and recorded every 10 minutes whether the participant adjusted one of the workstation features (e.g., chair, monitor).

**Figure 7.15** Prototype workstation showing seated and standing positions (Author)

The occurrence, not amount, of adjustment was recorded. During the experiment, researchers observed the participants and recorded any changes or adjustments they made to their workstations. Physiological indicators of stress were monitored by measuring epinephrine and free cortisol levels in urine samples collected at intervals during the study. Immediately following the 3 hour manuscript and clerical tasks activities, subjects filled out three questionnaires in the following order: Perceived Workstation Adjustability, Job Stress, and Postural Discomfort. This was immediately followed by the motivation measure consisting of a series of four geometric line puzzles (Cohen, 1980).

## D. Results

The general data analytic strategy was a 2 (Station type -- Prototype Workstation versus Standard Cubicle) X 2 (Training condition) analysis of variance (ANOVA). The results are organized into three major subsections:

Work Station Adjustability, Stress, and Postural Discomfort.

1. Workstation adjustability

The initial set of results focuses on adjustability of the work stations (see Table 7.9). Two overall indices of Perceived Adjustability provide inconsistent information. Magnitude Estimation ("Overall, how easy was the work station environment to adjust?" 1 = Extremely Difficult - 10 = Extremely Easy) indicated that training substantially increased perceived adjustability, F (1,79) = 6.85, p < .01 whereas both Station, F (1, 79) = 1.0 and their interaction, F (1, 79) = 1.01 were not significant. When participants were asked about particular components of the work stations, on average the Prototype Workstation was seen as significantly more readily adjustable than the Standard Cubicle, F (1,79) = 6.91, p < .01 whereas Training, F (1, 79) < 1.0 and the interaction of Training and Station were not significant, F (1, 79) = 2.99.

| Table 7.9 Perceived Adjustability of the Workstation | | Prototype Workstation | Standard Cubicle |
|---|---|---|---|
| **Training** | Magnitude Estimation | 7.53 | 7.60 |
| | Perceived Adjustability | 4.07 | 3.96 |
| | | | |
| **No Training** | Magnitude Estimation | 7.00 | 7.42 |
| | Perceived Adjustability | 4.22 | 3.73 |

The Results depicted in Table 7.9 are a composite for perceived adjustability across components held in common between the two workstations.

As can be readily seen in Table 7.10, Training increased perceived adjustability, especially for the monitor arm, main work surface and mouse tray on the Prototype Workstation and the File Trays and monitor position in the Traditional Cubicle.

Knowledge of adjustability was marginally greater for the Prototype Workstation relative to the standard cubicle, F (1, 79) = 2.94, p < .09. Training significantly enhanced knowledge about the adjustability of the workstations, F (1, 79)= 13.13, p <. 001, and this was particularly true for the Prototype Workstation, F (1,79) = 5.65, p < .001.

Objective adjustability of the two respective work stations was recorded every ten minutes as the participants worked. Table 7.11 depicts the overall results for the number of changes that occurred out of a possible 19 observation periods. The strong overall impact of greater adjustability of the Prototype Workstation is readily apparent where an average of 10.65 changes were made versus 7.68 for the traditional cubicle, F (1,79) = 14.08, p < .001. Neither Training, F (1,79) < 1.0 nor the interaction of Training and Station, F (1, 79) < 1.0 were significant.

## 2. Stress

Stress was measured with multiple methods including self-report (perceived stress), motivation performance, and physiology.

*Perceived stress.* Overall perceived stress showed a marginal main effect for Station, F(1, 78) = 3.04, p < .08 with the cubicle overall more stressful. Training had no main effect on perceived stress, F(1, 78) < 1.0. The interaction of station and training was significant, however, F(1, 78) = 3.84, p < .05, indicating that training reduced stress for the Prototype Workstation but not for the Standard Cubicle.

## 3. Motivation performance

Following completion of three hours of clerical work, individuals were given an index of motivation measured as persistence on challenging puzzles. Participants who worked on the Prototype Workstation but only when given training, persisted significantly more on the challenging puzzles, F(1,79) = 19.63, p < .0001.

| Table 7.10 Training Effects on Components Unique to each Workstation | | |
|---|---|---|
| **Prototype Workstation** | **Training** | **No Training** |
| **Light** | 4.25 | 4.32 |
| **Monitor arm** | 4.30 | 3.15** |
| **Keyboard tray** | 3.20 | 3.10 |
| **Work surface** | 4.26 | 2.80*** |
| **Mouse tray** | 4.00 | 2.63** |
| **Tack screen** | 4.30 | 3.95 |
| | | |
| **Standard Cubicle** | | |
| **Monitor tilt** | 4.25 | 3.67 |
| **Monitor position** | 4.15 | 3.22* |
| **File trays** | 4.45 | 3.44* |
| **Keyboard** | 4.00 | 2.93 |
| *p < .05  **p < .01  ***p < .001 | | |

It is important to note that when given no training, participants on the Prototype Workstation manifested the lowest levels of motivation. There is also a marginal main effect of station, $F(1, 79) = 3.56$, $p < .06$ and a highly significant effect of training, $F(1, 79) = 41.97$, $p < .0001$. People who were trained persisted nearly twice as much on the motivational index.

4. Physiological stress

When the physiological stress results were analyzed, we found

| Table 7.11 Objective Adjustability of the Workstation[a] | | |
|---|---|---|
| | **Prototype Workstation** | **Standard Cubicle** |
| **Training** | 10.45 | 7.95 |
| **No Training** | 10.85 | 7.40 |

a.    Number of intervals in which at least one adjustment was made during the 10 minute interval across the 3 hour work period. The possible range is from 0-19.

epinephrine was not affected directly by training, $F(1, 79) = 2.10$ nor by station, $F(1, 79) = 1.61$. There was, however, a significant interaction between these two factors, $F(1, 79) = 4.61$, $p < .05$. When participants were trained and worked on the Prototype Workstation, they experienced significantly less stress. Norepinephrine was unaffected by training, $F(1, 79) = 1.62$, station type, $F(1, 79) < 1.0$, nor their interaction, $F(1, 79) < 1.0$. Urinary cortisol was significantly affected by training, $F(1, 79) = 7.77$, $p < .01$ with participants who had been administered training having lower levels of cortisol. Neither station, $F(1, 79) < 1.0$ nor the interaction of station and training, $F(1, 79) = 1.98$ were significant.

5. Postural Discomfort

For the analyses, major body areas included head and neck, upper limbs including wrists and hands, leg and ankles, hip, and upper torso (back and chest). Upper body aggregated head and neck, upper limbs and torso. Lower body included leg and ankles and hip. In general few effects were found at the level of specific body areas.

For the upper body there were no effects of either training or work station type. For lower body, however, there was a significant benefit of training, $F(1,79) = 3.09$, $p < .01$. Those with training indicated significantly less postural discomfort, $M = 1.64$ in comparison to those without training, $M = 2.98$ at the end of the three hour working session. Statistical controls for initial levels of postural discomfort prior to the onset of the experiment had no impact on any of the postural discomfort findings. Much of the training effect appeared to derive from discomfort in the hips because training had large, significant impacts on both the left and right side of the hips.

Although there were no overall workstation effects on discomfort, the Prototype Workstation consistently produced lower discomfort ratings, and in the case of the torso (upper and lower back plus chest), significantly so.

## E. Conclusions

The findings of this laboratory study support the findings of field studies (O'Neill, 1998) which suggest that adjustability in the workplace has important effects on outcomes related to worker performance, motivation, postural discomfort and stress. This study also provides interesting insights regarding the importance of ergonomic training in the effective use of adjustments provided. Finally, as the first study to attempt to quantify the relationship between specific workstation design characteristics and physiological stress reactions, the experiment yielded some provocative findings on the effects of adjustability and training on stress-related health risks to workers.

### 1. Given adjustability, people exert control

Participants using the more highly adjustable prototype workstation adjusted it significantly more often than did participants working in the traditional cubicle, regardless of whether they had received ergonomic training.

### 2. Adjustability + training = reduced stress

Workers who received ergonomic training and used the prototype workstation had the lowest levels of epinephrine, and thus experienced significantly less physiological stress than participants in any of the other Experimental groups.

This indicates that greater opportunity for workstation adjustability, coupled with appropriate training about how to take advantage of that design adjustability, leads to lower stress. Without training, there appeared to be no difference, or even a slight increase in epinephrine levels (and stress), among workers who used the highly adjustable prototype when compared to those in the standard cubicle.

Although the level of adjustability had no effect on cortisol levels, this measure was significantly affected by training. Regardless of workstation adjustability, participants who received training had the lowest levels of cortisol.

### 3. Adjustability + training = enhanced motivation performance

The puzzle-solving test we used to assess performance motivation dramatically confirmed the importance of providing training along with adjustability. Participants who received training along with the high-adjustability prototype workstation showed the highest levels of motivational performance, while the prototype workstation users who had *not* been trained showed the lowest levels of motivational performance.

## F. Implications and Recommendations

We conclude that offering people control over their immediate work environment through providing adjustable workstations and the knowledge of how to properly use the adjustment options provided can lower stress and improve motivational performance in the workplace. However, providing office furniture and technology that increase the range of user adjustability will, in general, not lead to benefits unless the worker is instructed in the proper use of increased opportunities for adjustability. Options for adjustability without knowledge and practice in utilizing the options have little positive impact and in fact may cause unintended negative effects on stress and motivation.

In this study we attempted to simulate a high load work task representative of individual-based knowledge work process. Thus we did not consider other job types (such as highly routinized work in forms processing or customer service) nor did we assess the impact of these variables on team work process. We may want to consider these variables in future research. It seems likely however that if individual stress and motivation is impacted by the variables we considered, we would see some impacts on other types of jobs and work process styles, such as group work.

Business managers and professionals responsible for the provision of work environments may want to bias work environment design solutions towards those providing greater adjustability. However, the decision to invest in greater adjustability must be paired with a similar investment in ongoing training in the use of highly adjustable workstations in order to realize the full health and motivational performance benefits of this strategy. In fact, the use of highly adjustable workstations without training can actually reduce motivation performance.

The well established link between the physiological stress indicators used in this study, and important health consequences such as heart disease, provide an opportunity for business managers to think about the deployment of work environments in a strategic fashion within the organization. The impact of a coordinated program of training and work environment design could be integrated as part of a broader corporate strategy to control health costs and provide a healthier and more productive work environment for employees. The measurement of work environment design and training costs, coupled with assessment of the true financial and non-financial benefits of such a program could create a powerful business case that shares the benefits to employees and the organization.

# REFERENCES

Altman, R. and Rogoff, B. World views in psychology: Trait, interactional, organismic and transactional perspectives, In D. Stokols and I. Altman (Eds.), *Handbook of Environmental Psychology,* Vol. 1, 7-40, 1987.

Bader, G., Chang, R., and Bloom, A. *Measuring Team Performance: A Practical Guide to Tracking Team Success.* John Wiley & Sons, New York, 1999.

Baum, A. and Grunberg, N. Measurement of stress hormones. In S. Cohen, R.C. Kessler, and L. Gordon (Eds.), *Measuring stress,* 175-192, 1995.

Becker, F. D. *Office at work: Uncommon strategies that add value and improve performance.* Jossey-Bass, San Fransisco, 2004.

Brill, M., Margulis, S., Konar, E., and BOSTI. *Using office design to increase productivity.* Vols. 1 and 2. Workplace Design and Productivity, Buffalo, NY, 1984-1985.

Bruning, N. S. and Frew, D. R. Effects of exercise, relaxation, and management skills training on physiological stress indicators: a field experiment. *Journal of Applied Psychology,* 72(4), 515-521, 1987.

Campbell, D. R. and Stanley, J. C. *Experimental and Quasi-experimental Designs for Research.* Rand McNally, New York, 1966.

Caplan, R. D., Cobb, S., French, J. R. P., Van Harrison, R., and Pinneau Jr., S. R. *Job Demands and Worker Health.* U.S. Government Printing Office, Washington, D.C. 1975.

Carayon, P. Stressful jobs and non-stressful jobs: a cluster analysis of office jobs. *Ergonomics,* 37, 2, 311-323, 1994.

Cohen, S. New Approaches to Teams and Teamwork, In Galbraith, J. R. and Lawler, E. E. III (Eds.) *Organizing for the Future,* Jossey-Bass, San Francisco, 1993.

Coleman, C. *Interior Design Handbook of Professional Practice.* McGraw-Hill Companies Inc., New York, 2002.

Cooper, C. L. and Cartwright, S. Healthy mind: Healthy organization -- A proactive approach to occupational stress, *Human Relations,* 47(4), 455-471, 1994.

Dainoff M J. Occupational stress factors in visual display terminal (VDT) operation: a review of empirical research. *Behaviour and Information Technology* 1(2):141-176, 1982.

Davenport, T. *Thinking for a living*. Harvard Business School Press, Boston, 2005.

DeGraff, J. and Lawrence, K. A. *Creativity at Work: Developing the Right Practices to Make Innovation Happen*. University of Michigan Business School Management Series, John Wiley & Sons, New York, 2002.

Eckes, G.. *The Six Sigma revolution: How GE and others turned process into profits*, John Wiley & Sons, New York, 2001.

Edwards, J. R. and Cooper, C. L. The person-environment fit approach to stress: recurring problems and some suggested solutions. *Journal of Occupational Behavior*, 11, 293-307, 1990.

Evans, G.W. and Cohen, S. Environmental stress. In D. Stokols & I. Altman (Eds.), Handbook of environmental psychology (pp.571-610), John Wiley & Sons, New York, 1987.

Frese, M. Theoretical models of control and health. In Sauter, S. L., Hurrell Jr., J. J., and Cooper, C. L. (Eds.), *Job Control and Worker Health*. John Wiley & Sons, London, 1989.

Freney, R. *Pulse: The coming age of systems and machines inspired by living things*. Farrar, Straus and Giroux, 2006.

George, M. L. *Lean Six Sigma: Combining Six Sigma quality with Lean speed*. McGraw-Hill, New York, 2002.

Gladstein, D. Groups in context: a model of task group effectiveness. *Administrative Science Quarterly*, 29, 499-517, 1984.

Glass, D. C., Singer, J. E., and Pennebaker, J. W. Behavioral and physiological effects of uncontrollable environmental events. In D. Stokols (Ed.) *Perspectives on Environment and Behavior*, Plenum, New York, 1977.

Goodman, P. S. Impact of task and technology on group performance. In P. S. Goodman and Associates (Eds.), *Designing Effective Work Groups*. 120-167. Jossey-Bass, San Francisco, 1986.

Gordon, J. Work Teams: How far have they come? *Training*, 59-65, 1992.

Greider, W. *One World, Ready or Not*. Simon & Schuster, New York, 1997.

Grunberg, N. and Singer, J.E. Biochemical measurement. In J. Cacioppo and L. Tassinary (Eds.), *Principles of psycho-physiology.* 149-176, Cambridge Press, New York, 1990.

Hahn, A. Facility performance and serviceability from a facility manager's viewpoint. In G. Davis and F. Ventre (Eds.), ASTM STP 1029, *Performance of Buildings and Serviceability of Facilities*, ASTM, Philadelphia, 1990.

Hedge, A. Job stress, job satisfaction, and work-related illness in offices, *Proceedings of the Human Factors Society 32nd Annual Meeting*, 777-779, 1988.

Herman Miller, Inc. *Research Report: A test of team and temporal work spaces.* HMI Product Research Group, Holland, MI, 1994.

Herman Miller, Inc. *Research Report: Liaison User Test*. Herman Miller Product Research Group, Holland, MI, 1993.

Jackson, S. Does job control control job stress? In S.L. Sauter, J. Hauter, and C.L. Cooper (Eds.), *Job control and worker health.* 25-54, John Wiley & Sons, New York, 1989.

Johansen, R. and Swigert, R. *Upsizing the Individual in the Downsized Organization.* Addison-Wesley, New York, 1994.

Johansen, R., Sibbet, D., Benson, S., Martin, A., Mittman, R., and Saffo, P. *Leading Business Teams*. Addison-Wesley, Reading, MA, 1991.

Johnson, J. V., Stewart, W., Hall, E. M., Freelund, P. and Theorell, T. Long-term psychosocial work environment and cardiovascular mortality among Swedish men, *American Journal of Public Health*, 86(3), 324-331, 1996.

Jones, S. D., Buerkle, M., Hall, A., Rupp, L., and Matt, G. Work group performance measurement and feedback, *Group and Organization Management*, 18, 3, 269-291, 1993.

Kaplan, R. S. and Norton, D. P. *The Balanced Scorecard*. Harvard Business School Press, Boston, MA, 1996.

Karasek, R. A., Gardell, B., and Lindell, J. Work and non-work correlates of illness and behaviour in male and female Swedish white-collar workers. *Journal of Occupational Behavior*, 8, 187-207, 1987.

Karasek, R.A. and Theorell, T. *Healthy work: stress, productivity, and the reconstruction of working life*. Basic Books, New York, 1990.

Karasek, R. A. Job demands, job decision latitude, and mental strains: Implications for job redesign. *Administrative Science Quarterly*, 24, 285-308, 1979.

Kasl, S. V. An epidemiological perspective on the role of control in health. In S. L. Sauter, J. J. Hurrell Jr., and C.L. Cooper (Eds.), *Job Control and Worker Health.* John Wiley and Sons, London, 1989.

Keppel, G. and Wickens, D. *Design and Analysis: A Researcher's Handbook, Third Edition*, Prentice-Hall, Englewood Cliffs, NJ, 2004.

Land, G. and Jarman, B. *Breakpoint and Beyond: Mastering the Future Today.* HarperBusiness, New York, 1993.

Landsbergis, P. A. Occupational stress among health care workers: a test of the job demands-control model. *Journal of Organizational Behavior*, 9(3), 217-239, 1988.

Lawler, E. E., Mohrman, S., and Ledford, Jr., G. *Creating High Performance Organizations.* Jossey-Bass Publishers, San Francisco, CA, 1995.

Light, P. C. *The Four Pillars of High Performance.* McGraw-Hill, New York, 2005.

Nadler, D. A. *Feedback and Organizational Development: Using Data-Based Methods.* Addison-Wesley, Reading, MA, 1977.

O'Leary-Kelly, A. M., Martocchio, J. J., and Frink, D. D. A review of the influence of group goals on group performance. *Academy of Management Journal*, 37, 1285-1301, 1994.

O'Neill, M. J. Job type, workstation design, and effective work. *In Proceedings of the 1995 Human Factors and Ergonomics Society 39th Annual Meeting*, 819-823, 1995a.

O'Neill, M. J. and Carayon, P. The relationship between privacy, control and stress responses in office workers. *In Proceedings of the Human Factors and Ergonomics Society 37th Annual Meeting*, 479-483, 1993.

O'Neill, M. J. Evaluation of a conceptual model of architectural legibility. *Environment and Behavior*, 23(3), 259-284, 1991a.

O'Neill, M. J. A biologically based model of spatial cognition and wayfinding. *Journal of Environmental Psychology*, 11, 299-320, 1991b.

O'Neill, M. J. Effects of familiarity and plan complexity on wayfinding in simulated buildings. *Journal of Environmental Psychology*, 12, 319-327, 1992.

O'Neill, M. J. Job Type, Workstation Design and Worker Stress and Health. *In Proceedings of Work, Stress and Health '95: Creating Healthier Workplaces*, Washington, D.C. 1995b.

O'Neill, M. J. Work space adjustability, storage and enclosure as predictors of employee reactions and performance. *Environment and Behavior*, 26, 4, 504-526, 1994.

O'Neill, M. J. *Ergonomic Design for Organizational Effectiveness*. Lewis Publishers, Boca Raton, FL, 1998.

O'Neill, M. J. and Evans, G. Effects of Workstation Adjustability and Training on Stress and Motivational Performance. *In the Proceedings of the 2000 EDRA Conference*, Los Angeles, CA, 2000.

O'Neill, M. J. and Duvall, C. A Six Sigma quality approach to workplace evaluation. *Journal of Facilities Management*, Vol. 3, Number 3, Henry Stewart Publications, London, 2005.

O'Neill, M. J. Theory and research in design of "You are Here" maps. In H. J. Zwaga, T. Boersema, and H. Hoonhout (Eds.), *Visual information for everyday use: Design and research perspectives*. 225-238, Taylor & Francis, Philadelphia, PA, 1999.

Paciuk, M. The role of personal control of the environment in thermal comfort and satisfaction at the workplace. In R. Selby, K. Anthony, J. Choi, B. Orland (Eds.), *Proceeding of the EDRA 1990 Conference*, 303-312, 1990.

Pepper, S. C. *World hypotheses: a study in evidence*. University of California Press, Berkeley, 1942.

Perrewe, P. L. and Ganster, D. C. The impact of job demands and behavioural control on experienced job stress. *Journal of Organizational Behavior*, 10, 213-229, 1989.

Reese, H. W., and Overton, W. F. Models of development and theories of development. In L. R. Goulet and P. B. Baltes (Eds.), *Life-span Developmental Psychology: Theory and research*, 115-145, New York; Academic, 1970.

Rothschild, M. *Bionomics: Economy as Business Ecosystem*. Beard Books, New York, 1990.

Sainfort, P. Job design predictors of stress in automated offices. *Behavior and Information Technology*, 9, 1, 3-16, 1990.

Salustri, J. Discussion of operating performance. In G. Davis and F. Ventre (Eds), ASTM STP 1029, *Performance of Buildings and Serviceability of Facilities.* ASTM, Philadelphia, 1990.

Sauter, S. L., Hurrell, J., and Cooper, C. L. (Eds.). *Job control and worker health.* John Wiley & Sons, New York, 1989.

Scalet, E. *VDT Health and Safety: Issues and Solutions.* ErgoSyst Associates Publication, KS, 1987.

Semmer, N. and Frese, M. *Control at work as moderator of the effect of stress at work on psychosomatic complaints: A longitudinal study with objective measurement.* Department of Psychology, University of Bern, Switzerland. Manuscript, 1988.

Shadish, W. R., Cook, T. D., and Campbell, D. T. *Experimental and Quasi-Experimental Designs For Generalized Causal Inference.* Houghton Mifflin, New York, 2001.

Smith, M. J. and Sainfort, P. A balance theory of job design for stress reduction. *International Journal of Industrial Ergonomics*, 4, 67-79, 1989.

Sundstrom, E. *Work Places.* Cambridge University Press, New York, 1986.

Sundstrom, E. and Altman, I. Physical environments and work group effectiveness. *Research and Organizational Behavior,* Vol. 11. 175-209, 1989.

Sundstrom, E., De Meuse, K. P., and Futrell, D. Work teams: applications and effectiveness. *American Psychologist,* 120-133, February, 1990.

Taylor, F. *The Principles of Scientific Management.* Norton, New York, 1911, 1966.

Taylor, J. C. and Felton, D. F. *Performance by Design.* Prentice Hall, New Jersey, 1993.

Uehata, T. Long working hours and occupational stress-related cardiovascular attacks among middle-aged workers in Japan. *Journal of Human Ergology,* 20, 147-153, 1991.

Visher, J. C. *Workspace strategies: Environment as a tool for work.* Chapman and Hall, New York, 1996.

Wheeler, D.J. and Chambers, D.S. *Understanding Statistical Process Control, Second Edition.* SPC Press, Inc., 1992.

Wheeler, D.J. *Advanced Topics in Statistical Process Control: The Power of Shewhart Charts.* SPC Press, Inc., 1995.

Zimmer, L. and Cornell, P. An examination of flexible work spaces in the open office. *Proceedings of the HFS 34th Annual Meeting,* 890-894, 1990.